T0129652

Copernicus Books

Sparking Curiosity and Explaining the World

Drawing inspiration from their Renaissance namesake, Copernicus books revolve around scientific curiosity and discovery. Authored by experts from around the world, our books strive to break down barriers and make scientific knowledge more accessible to the public, tackling modern concepts and technologies in a nontechnical and engaging way. Copernicus books are always written with the lay reader in mind, offering introductory forays into different fields to show how the world of science is transforming our daily lives. From astronomy to medicine, business to biology, you will find herein an enriching collection of literature that answers your questions and inspires you to ask even more.

Amandine Aftalion

Be a Champion

40 Facts You Didn't Know About Sports and Science

Translated by the author from *Why do you lean in a bend? Sports explained by science in 40 questions*

Amandine Aftalion
CNRS
Paris, France

Characters were drawn by Estelle Chauvard

ISSN 2731-8982 ISSN 2731-8990 (electronic)
Copernicus Books
ISBN 978-3-031-54081-3 ISBN 978-3-031-54082-0 (eBook)
https://doi.org/10.1007/978-3-031-54082-0

English translation of the original French edition "Pourquoi est-on Penche dans les Virages," published by CNRS Editions, Paris, 2023

© The Editor(s) (if applicable) and The Author(s), under exclusive license to Springer Nature Switzerland AG 2024
This work is subject to copyright. All rights are solely and exclusively licensed by the Publisher, whether the whole or part of the material is concerned, specifically the rights of reprinting, reuse of illustrations, recitation, broadcasting, reproduction on microfilms or in any other physical way, and transmission or information storage and retrieval, electronic adaptation, computer software, or by similar or dissimilar methodology now known or hereafter developed.
The use of general descriptive names, registered names, trademarks, service marks, etc. in this publication does not imply, even in the absence of a specific statement, that such names are exempt from the relevant protective laws and regulations and therefore free for general use.
The publisher, the authors, and the editors are safe to assume that the advice and information in this book are believed to be true and accurate at the date of publication. Neither the publisher nor the authors or the editors give a warranty, expressed or implied, with respect to the material contained herein or for any errors or omissions that may have been made. The publisher remains neutral with regard to jurisdictional claims in published maps and institutional affiliations.

This Copernicus imprint is published by the registered company Springer Nature Switzerland AG
The registered company address is: Gewerbestrasse 11, 6330 Cham, Switzerland

Paper in this product is recyclable.

TABLE OF CONTENTS

LIST OF BOXES

Preface

There is something fascinating about speed. We all dream of great speed, and we all know that speed is the distance covered in time. But how can you adjust your speed in the best way to go as fast as possible at all times?

Sports is an activity involving physical exertion in which an individual or a team competes against one another. Therefore, it involves movement and performance. The natural questions that arise immediately are: how to be the best, the quickest, the strongest, how to limit exertion and optimize effort, and how to be a champion?

Our choices are governed by the goal to make the best decision and to develop the optimal strategy. How can we measure, how can we decide what is "best"? This measure of the best can be described by an objective which varies from one sport to another. This requires optimisation theory with an economic formulation of cost and benefit, whether you are interested in minimising time, trajectory, energy expense, effort, or the resistance of a material. Mathematical optimisation makes it possible to combine all the constraints to determine the best strategy.

The world of sports is filled with numbers: records that appear at the end of each competition that measure time, length, and all sorts of sensors provide us with position, speed, force, power, energy. All these numbers are called data. They are used together with photos and videos to analyse movement and to assist referees and coaches. Data is not sufficient to help us optimize sports performance; equations and deterministic formulation provide more of an insightful understanding of phenomena.

Be a Champion presents these and many other uses of mathematics and physics that can help in sports strategy and performance and allows us to understand and discover sports in a new way.

Amandine Aftalion

Paris

CHAPTER 1

RUNNING

WHY DO YOU LEAN IN A BEND?

Whether on a bike, on a motorcycle, or on horseback, whether skiing, or ice skating, sportsmen and women all lean in a bend. And what about runners? Runners lean too, but less than skaters. Why? Simply because they don't go as fast!

Usain Bolt leaning in a bend. (Photo by Erik van Leeuwen)

© The Author(s), under exclusive license to Springer Nature Switzerland AG 2024
A. Aftalion, *Be a Champion*, Copernicus Books,
https://doi.org/10.1007/978-3-031-54082-0_1

What happens in a train turning left? Where are you pulled to? To the right, of course. You are thrown out of the bend by the centrifugal force, and you have to hold back or exert a force to straighten out. The natural reaction is to lean to the left to keep your balance. It is the same thing on a merry-go-round: when it starts moving, you have to hold back to avoid being thrown off, and it is easy to see that the wires of fast-moving merry-go-rounds lean outwards.

What about a runner in a bend? There are no wires to hold him back, so, to take a left turn, he leans to the left to go against the centrifugal force. It is not easy to feel this force when you are running. It is proportional to the square of your speed, i.e., your speed multiplied by itself. So the faster you go, on a bike, on a motorcycle, ice skating or skiing, as

Skater leaning into a bend with his hand on the ice.

examples, the more you lean. Speed skaters and skiers even put their hand on the ground because they lean so far down. We know how to calculate the value of the centrifugal force, which is:

$$m\frac{v^2}{R}$$

where v is the velocity, R the circle radius and m is the athlete's mass. If the curve is not exactly circular, R is the radius of the circle that is closest to the curve. In the case of a straight line, the radius is infinite, so the inverse of this radius is zero; there is no centrifugal force!

On the other hand, on a track with a semicircle of radius $R = 36.5$ m, the centrifugal force per unit of mass is of the order of 3 newton per kilogram for a speed of 10 m/s (that of sprinters). This is to be compared with a weight per unit mass of $g = 9.81$ newton per kilogram. The centrifugal force exerted on a sprinter therefore represents a third of his weight, which is quite huge! Even if it is not obvious that sprinters are leaning, the effect of this force must be taken into account when considering performance.

The angle of leaning is of the order of 15° for a runner: it is calculated using its tangent, which is equal to

$$\tan\theta = \frac{v^2}{Rg}$$

where θ is the angle of inclination. A runner who has to take a sharp turn uphill will prefer to take it on the outside in order to reduce the centrifugal force he has to fight against, even if this means covering a slightly greater distance. In the same way, when coming out of a bend onto a straight path, one prefers to avoid a sudden change in centrifugal force and anticipates this by straightening up. Motorcyclists know this well. In anticipation of exiting the bend, they straighten the motorbike, which is hard on the arms.

WHY DO YOU RUN WITH FOLDED ARMS RATHER THAN STRAIGHT ARMS?

While skaters glide on the ice, runners jump from one leg to the other. Every time a runner changes his supporting leg, his hips rotate. When his left foot touches the ground, he pushes off. This propulsive force turns the hips to the left and propels the right leg forward in the air. To balance the body and avoid falling over, especially when running fast, the runner rotates his shoulders in the opposite direction using his left arm in front. The image shows that there is a strong torsion between the hips and the shoulders, with the arms serving to stabilise the body.

Runner with folded arms. Shoulders and hips rotate in opposite directions.

When the right foot hits the ground, the situation reverses and the arms swing in the opposite direction to induce what is known as a "corrective torque", which again helps to keep the body facing forward.

In any walking or running movement, you are stabilised by the movement of the arms which is opposite to that of the legs.

Keeping the arms bent rather than straight helps with the swinging movement. As with a swinging pendulum, the longer the wire, the harder it is to swing, and the shorter the wire, the easier it is to swing. The arms swing easier when they are bent, because they are then shorter.

Is it useful to pull far back with the arms to try and increase the size of the stride? The size of your stride depends on your height, weight, centre of gravity, leg length and joint flexibility. We all have an optimal stride that corresponds to a minimum energy cost. So if you want to perform, there is no point in trying to change your stride size from the one which is natural to you, by pulling back on your arms for example, as this could lead to exhaustion.

WHY DOES A SPRINTER SLOW DOWN BEFORE THE FINISH LINE?

In the most well-known sprint event, the 100 metres, athletes cross the finish line by decelerating! In fact, they slow down after 60 or 70 metres, after approximately 2/3 of the race. The 200 metres and 400 metres are also run by decelerating after the initial acceleration. It is only for distances above 1,500 metres that one accelerates at the end of the race. This is the best way to manage the energy and the effort, which cannot be maintained at a maximum throughout the race, even if this is the impression athletes give. So the maximum effort has to be produced at the start and then the deceleration has to be as small as possible. This phenomenon was very clearly exhibited in Usain Bolt's world record race in Berlin in 2009:

Distance in metres	Time in seconds	Δt in seconds	Speed in m/s
10	1.89	1.89	5.29
20	2.88	0.99	10.10
30	3.78	0.9	11.11
40	4.64	0.86	11.63
50	5.47	0.83	12.05
60	6.29	0.82	12.20
70	7.10	0.81	12.35
80	7.92	0.82	12.20
90	8.75	0.83	12.05
100	9.58	0.83	12.05

The second column of this table gives his time as a function of the distance covered. The third column gives the time taken to cover 10 metres, Δt, which is the difference between two consecutive lines in the time column. The last column is the average speed every 10 metres, so the distance, 10 metres, divided by the time, Δt.

If one plots the third column of the table as a function of distance, one can see that the time to cover 10 m is large at the beginning and then decreases, as the runner accelerates, but above all that this time decreases until 70 m of running and then increases again.

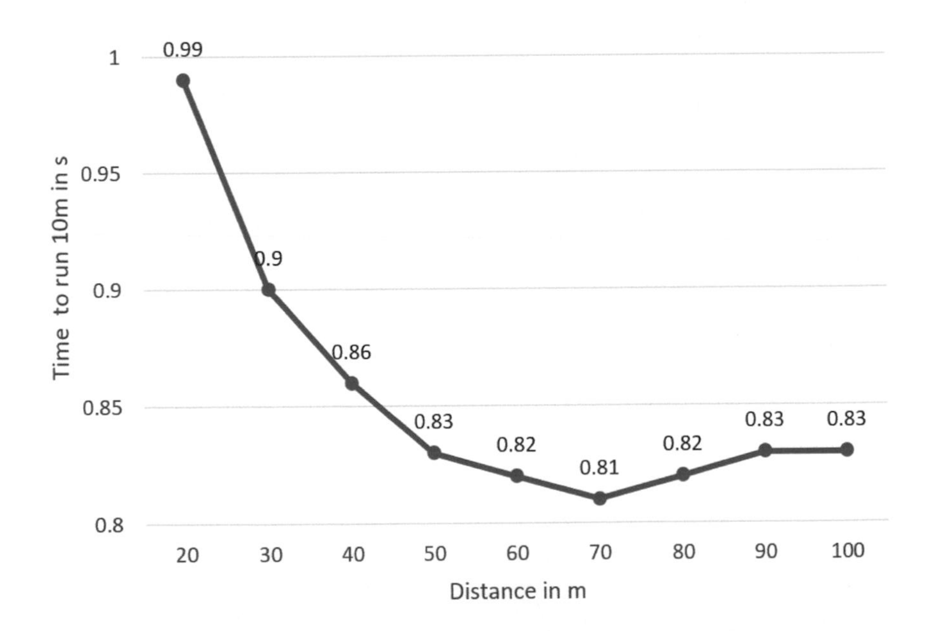

Time to run 10 m vs distance for Usain Bolt in Berlin, 2009.

This is even more obvious if you look at the fourth column, which gives the speed at the last 10 metres as a function of distance. The slowdown measured from 70 metres upwards is clear!

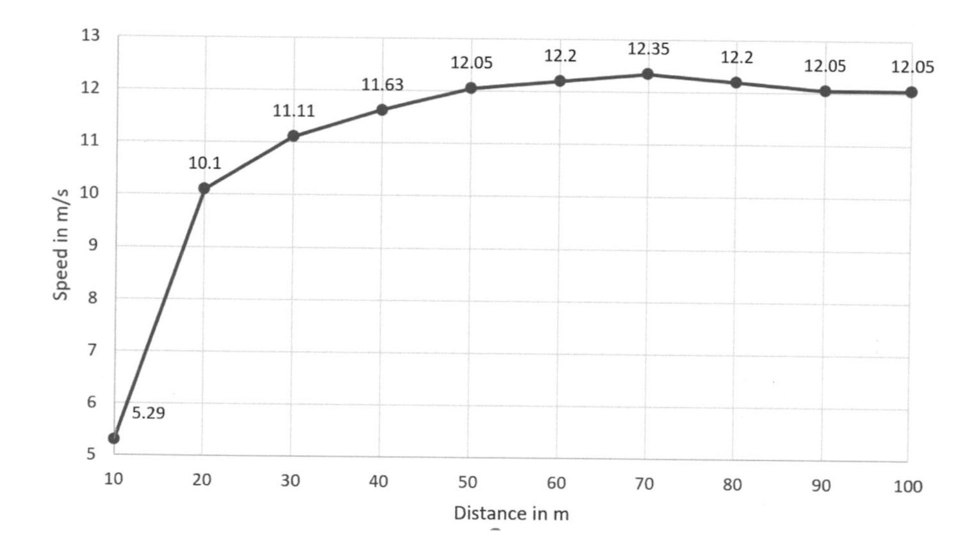

Speed to run 10 m vs distance for Usain Bolt in Berlin, 2009.

If one looks at the races of other champions, one would observe exactly the same phenomenon, sometimes greater deceleration occurs as early as 60 metres.

Why doesn't the human body optimise its resources in the same way for different distances? Mathematicians have produced models that help us to understand this. They have used data from champions and are able to simulate races in which virtual runners behave exactly the same way as real athletes. The specifics for short races is that the maximum flow of oxygen transformed into energy is not reached. This flow increases during the race without reaching the runner's maximum value. Sprinting is essentially "anaerobic", meaning that the energy resources used are not based on oxygen but on the energy stored in the body. If this stock was very large, the athlete could run at maximum speed throughout the race without slowing down at the end. Since this anaerobic contribution is limited, the best strategy is to start at full speed to accelerate immediately as much as possible and reach peak speed as quickly as possible. The athlete is unable to maintain his maximum propulsion force over the entire distance. This propulsion force therefore decreases and so does the speed. Calculations show that starting slower and accelerating throughout would be less profitable. The strategy is the same from the 100 m to the 400 m, which are races where the VO_2, the oxygen flow, increases, without reaching a plateau.

The equations used to describe running are based on physical laws, such as:

- conservation of energy: the energy used by the athlete to produce a propulsive force comes from two contributions: aerobic energy, due to respiration (in particular the VO_2, the flow of oxygen transformed into energy) and anaerobic energy, which does not depend on oxygen but on phosphocreatine and glucose.

- Newton's second law of mechanics: an equation that relates motion to the forces involved. More precisely, the variation in speed is equal to the propulsive force minus friction.
- theory of motor control which limits variations in propulsion force. For example, when you vary your speed, and decide to stop, it cannot be instantaneous because there is a reaction time between the decision in the brain and the action in the muscle.

Based on these laws, researchers establish and solve equations that enable them to assess speed, propulsion force, energy and motor control at any given time, and to gain a better understanding of the physiological phenomena involved, racing strategies, and performance.

Aerobic and anaerobic energy

Muscles need energy. It is obtained by converting ADP, adenosine diphosphate, into ATP, adenosine triphosphate. ATP is present in muscles in very small quantities and is used up in the first few seconds of exercise. In order to produce it, the body has to trigger one of three types of reaction, depending on the intensity and duration of effort:

1. ***ATP-CrP*** : ATP is restored by a molecule called phosphorylcreatine. This system dominates during exercise of very short duration and very high intensity (such as a sprint).

2. ***glycolysis*** : ATP concentration is maintained by using glucose as an energy substrate and by producing lactate during exercise. This system is dominant for exercises lasting less than a minute and of high intensity (such as the 400 m in athletics or the 100 m in swimming).
3. ***Aerobic energy*** : It is ubiquitous, but becomes dominant for ATP synthesis for exercises of medium-to-long duration and medium-to-low intensity. It involves the oxidation of carbohydrates and fats, which releases energy according to the formula $C_6H_{12}0_6 + 36\ Pi + 36\ ADP + 60_2 \rightarrow 6H_20 + 6CO_2 + 36\ ATP$ +2870 J.
   One mole of oxygen corresponds to 22.4 litres of pure oxygen, so in the formula 60_2 means 6 moles and corresponds to 134 litres of oxygen. An oxygen flow of 1 litre per kg per second produces 2 870/134= 21.4 joule per kg and second.

The VO_2 corresponds to the volume of oxygen that mitochondria, the organelles that supply our body's cells with the energy they need, can absorb per unit of time, and which is used in the oxidation reaction of fats and carbohydrates. The higher this value, the better the athlete performs, because he can produce more energy for his muscles per unit of time.

WHY SHOULDN'T YOU START TOO FAST IN AN ENDURANCE RACE?

"Slow and steady wins the race" as the fable says. Rushing is bad and it is quite true for endurance racing. If you start out too quickly, it is well known that you will have trouble getting to the finish and, above all, producing the decisive final acceleration. On the other hand, if you start out too slowly, you will certainly have reserves, but you'll never manage to make up for lost time. It is therefore very important to find the right acceleration speed at the start of a long race in order to be able to produce the final sprint. To do this, mathematicians have established models, as in the case of sprints, based on the conservation of energy, Newton's second law, and the theory of motor control. The difference between a sprint and an endurance race is that in the latter, a maximum flow of oxygen is maintained over a large part of the race, decreasing at the end, whereas in a sprint, the oxygen flow increases with the race.

A long run can be divided into three stages:

- An acceleration stage to reach peak speed which is higher than the cruising speed. The aim is to reach this speed as swiftly as possible. This acceleration phase is run "anaerobically", but its role is to launch the aerobic cycle, so that the flow of oxygen converted into energy reaches a plateau as quickly as possible. For this to happen, it is important to increase heart rate and accelerate hard. Well-trained athletes reach a peak speed above cruising speed, then slow down to cruising speed when the maximum oxygen flow is reached.

- An intermediate phase based essentially on aerobic energy. The greater the flow of oxygen converted into energy, the greater the cruising speed. This speed is not exactly constant if there are bends or hills. It can be useful to vary your speed slightly around your average cruising speed.
- The final sprint phase, which relies on residual anaerobic energy. When approximately a third of the initial anaerobic reserves remain, the aerobic mechanism decreases in intensity. This is the body's feedback mechanism and is the moment to accelerate. If the runner has accelerated too much at the start of the race, he won't have enough anaerobic reserves left for the final phase.

Over long distances, some champions make strategic accelerations in the middle of the race. In this way, they force their rivals to follow them as they accelerate, thereby expending anaerobic energy in the middle of the race. The rivals often have a lower anaerobic reserve, and therefore struggle to finish, because at the end of the race they no longer have sufficient residual anaerobic energy to produce the final acceleration.

The next figure shows an example of a runner who started too quickly (black) compared with the ideal race (red) and one who started too slowly (dotted line). The runner tracked in black sets off with a high propulsion force, which he maintains for 300 m, but then runs more slowly and is slower over all. The dotted runner starts off slower (his propulsion force was blocked at the beginning); his cruising speed will be higher, but it is too late. The dotted runner can't catch up and the black runner cannot accelerate enough at the end.

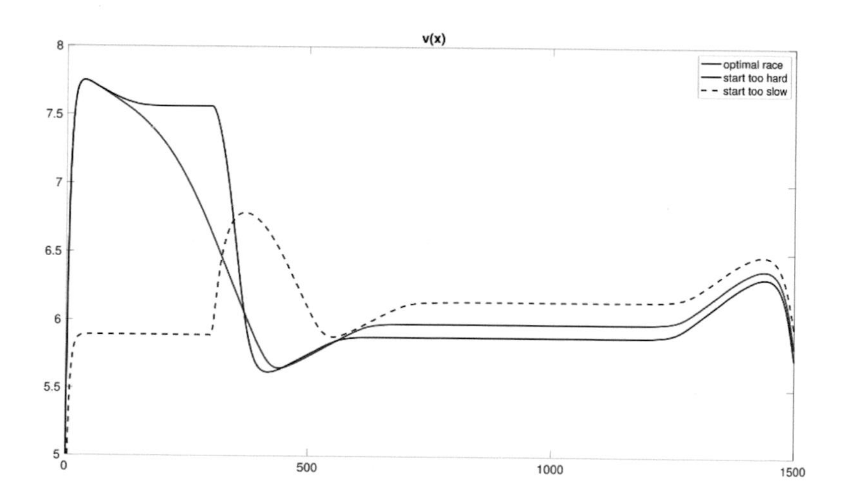

Speed on a 1500 m. The red curve shows the optimal speed. The black curve depicts a start that is too quick with a high propulsive force maintained on the first 300 m. Dashed curve, start too slow with a reduced propulsive force on the first 300 m.

A good tactic: following someone at your own pace saves energy and therefore time. It is not just a question of protecting yourself from the wind, as cyclists do, there is a psychological effect, because you don't have to think about your pace and you can keep going for longer. But you still need to set the right pace.

WHY ARE DEPARTURES STAGGERED ON AN ATHLETIC TRACK?

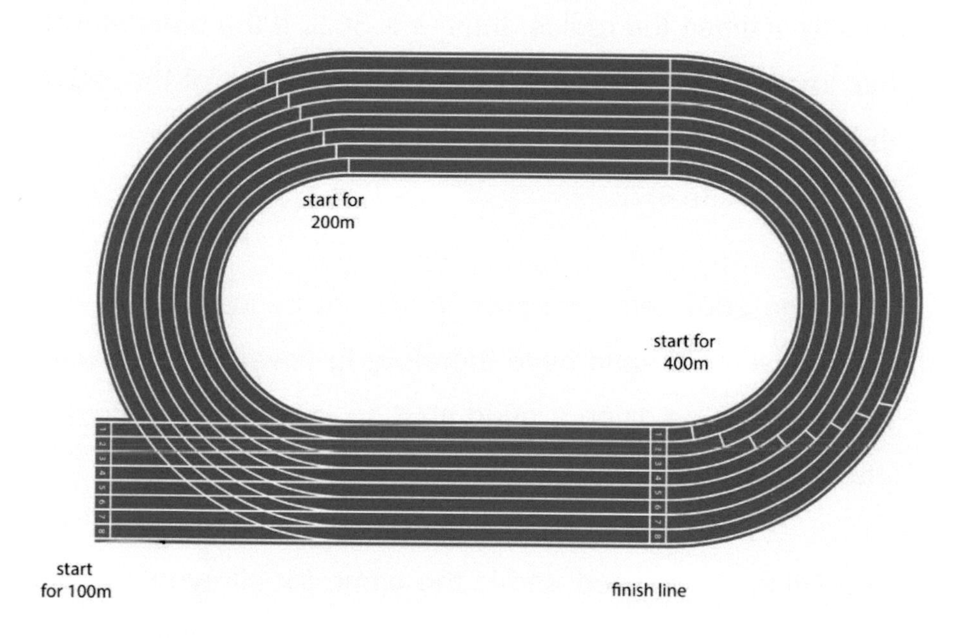

An athletic track. The 100m is run straight. The departures for the 200m and 400m are staggered.

The 100 metres is always run in a straight line, even if the straight line in a track is shorter, a specific zone extends the track on the side of the finish line to run 100 metres. But what exactly is the length of the straight line?

The official length of an athletics track is 400 m, measured 30 cm from the edge of the track. The width of standard stadiums is 73 m (historically imposed to contain a football pitch). Standard tracks are made up of two straight lines and two semi-circles. Since the diameter of the circle is 73 metres, its radius is 36.5 metres. It is therefore possible to calculate the length covered in a bend, which is proportional to the radius: it is exactly π times the radius, thus $\pi \times 36.5$. If the perimeter of the track is 400 metres, 30 cm from the edge of the track, the straights measure 84.39 metres each since it is $(400 - 2\pi \times 36.8)/2$.

From the 200 metre onwards, there are bends. Those who aren't in the first lane have therefore to cover more ground; they have to be given a head start so that everyone covers the same distance. How are these start lines defined?

The finish line is fixed and is the same for all events. Each lane is 1.22 m wide. For a 400 m race, i.e., a complete lap, the distance between the start lines of two successive lanes is therefore $1.22 \times 2\pi$, that is 7.66 m (except between the first and second lane where it is $1.12 \times 2\pi$, that is 7.04 m). For a 200 m, it is less than half. And for an 800 m race, runners only have to stay in their lane for 110 metres after which they break for position to the inside. This explains the staggered white start lines around the stadium.

Races have been run counter-clockwise since 1913. The reason is not scientifically established, but some studies in the desert have shown that a majority of people left alone to find their way tend to turn left. It is thought that this is linked to the fact that the heart is on the left. In any case, in the absence of wind, it has been measured that performance is better this way.

WHY DOES RUNNING BEHIND SOMEONE ALLOW YOU TO IMPROVE PERFORMANCE?

We are all familiar with the opposition between athletics as a sport of confrontation and a taste for records. A question that was recently revived when Kenyan Eliud Kipchoge successfully attempted to run the marathon in under 2 hours (1h 59 min 40 s) on October 12, 2019. He ran "alone", with the best possible equipment, on flat ground, and behind a car that gave him the pace. Given these specific conditions, the record was not officially recognised, but it does show that running behind someone can improve performance. It is not due to the aerodynamic effect. The reason: you do not have to think about your pace, which saves you energy when running. It has been measured that an athlete running around a track (400 m) behind someone can gain up to one second compared with running solo in the race. Additionally, in a 200 m or 400 m race, there is a disadvantage running in the outside lanes because you are running "blind", which is penalising. But in the inside lanes, the centrifugal force is greater and slows down the athlete.

By modelling the race, it is possible to understand the correlated effects of centrifugal force and the positive psychological effects generated by an athlete in front. The start of a 200 m race takes place in the bend. The psychological effect of having someone in front must be taken into

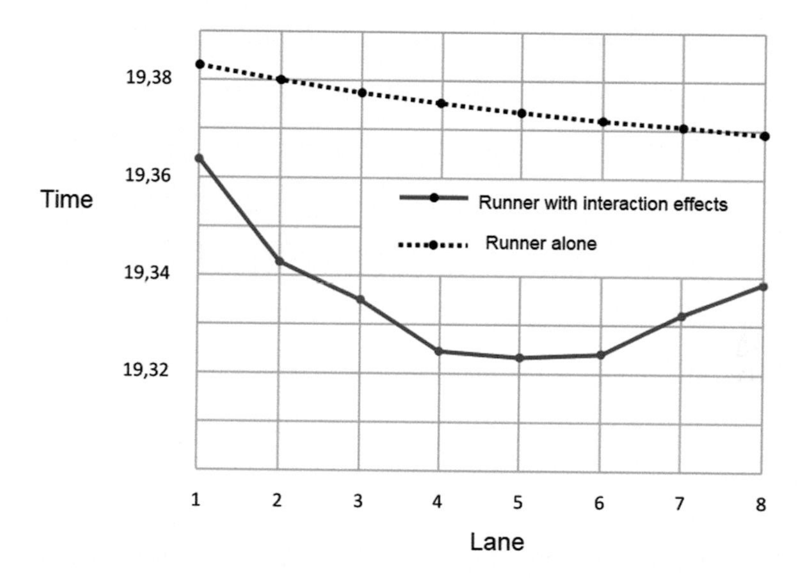

Time for 200 m vs lane. The black curve shows that for someone running alone, that is with the centrifugal force only, time decreases with lane number. The red curve shows the interaction effect when there is a runner on the next lane. The best lanes are in the centre: 4, 5, 6.

account in the model, but not in the case immediately after you have been overtaken because there is a reaction time once you are overtaken. This model can be used to study the time it takes a champion to run a 200 m race depending on in which lane they are running. If you run alone, the more outside the lane, the better the time (see the figure, black curve). The model can include the positive effect of having someone in the neighbouring lane. If you run with a competitor in the next lane, your performance is best for lanes 4 to 6, where the effect of centrifugal force and the psychological effect combine (see the figure, red curve). Next, lanes 7 and 3, then 8 and 2 are respectively equivalent, but lane 1 is heavily penalised. Note that in competitions until May 22, 2023, the best runners were allocated to lanes 3 to 6 by drawing lots, then the next two were allocated to lanes 7 and 8 by drawing lots and finally the last two were allocated to lanes 1 and 2. Since May 2023, the lane allocation depends on the length of the race, but the worst runners are still allocated to lanes 1 and 2.

So the best runners get the most favourable lanes and the worst performers get placed in the least favourable lanes. Therefore, there is little hope of seeing a winner in the first two lanes!

WHY ARE ATHLETIC TRACKS WITH LONGER STRAIGHTS WORSE FOR PERFORMANCE?

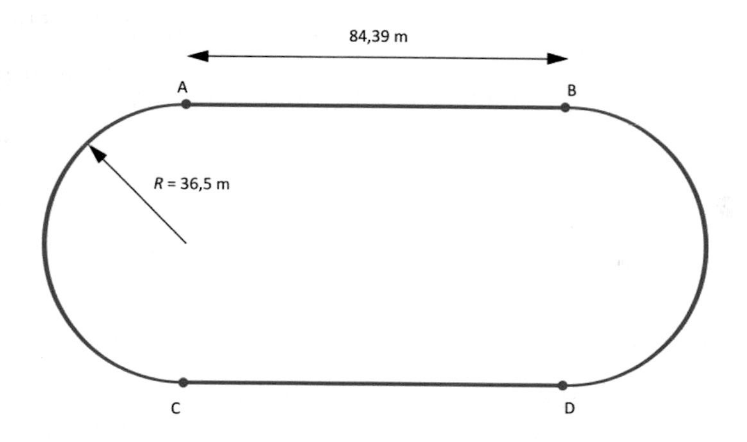

Standard track.

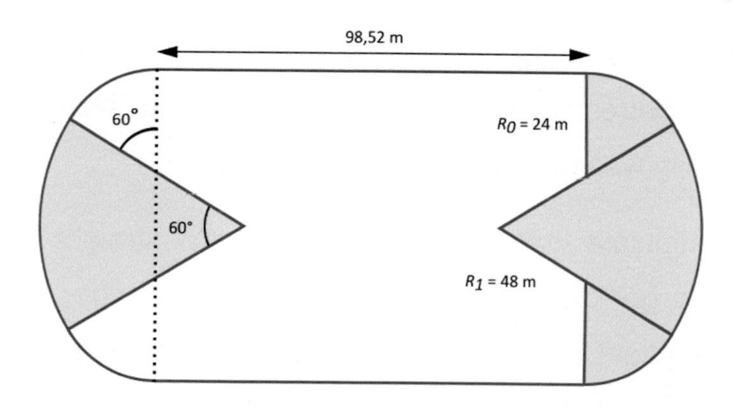

Double-bend track.

Three types of tracks are approved by the international athletics federation (World Athletics): the so-called "standard" tracks, made up of straight lines and semi-circles, and the so-called "double bend" tracks, whose curves are made up of arcs of several circles. The latter were introduced to include a football or rugby pitch and therefore have a much longer straight line and have bends with a small radius of curvature, even if the bend doesn't look that tight.

Double bend tracks are much less favourable to making record attempts, particularly in the inner lanes where the small radius of curvature (24 m) induces a very high centrifugal force which slows the runner down, even if he then benefits from a longer straight line.

A mathematical calculation can be used to determine the closed track that would give the best performance: it would be a circular track with a 63 m radius, i.e., a diameter of 126 m, which would obviously have significant disadvantages in terms of visibility for spectators, given that current stadiums are 73 metres wide. In the 1960s in France, however, the possibility was mentioned of building one.

If we stipulate that the track must contain straight lines, in order to get closer to a 100 m in a straight line, then the optimum track is made up of two straight lines and two semi-circles. But the performance calculation tells us that the shorter the straight, the better the time. Performance over 200 metres doesn't change whether you are on a straight line or on a curve with a straight line of less than 60 metres; but for a straight line of 60 metres, you gain 4 hundredths over 200 m compared with a straight line of 84 m 39, but above all you reduce the gap between the extreme lanes from 8 to 2 hundredths.

For the 10,000 m, shortening the stadium straight to 60 metres would save between 5 and 10 seconds. The shorter the straight, the wider the radius of the bend, and therefore the better the performance. But a stadium with a straight line of 60 metres would be a little short for the javelin throw and would draw spectators away from the arena.

Let us now look at the effect of centrifugal force on performance. It slows down the runner's effort. At the start of a sprint, in the acceleration phase, you don't necessarily notice it but in a longer race, with alternating straights and bends, you can clearly see that the runner's speed decreases as he goes round a bend. No matter how hard you try no matter how hard you push, you will always run slower in a bend than in a straight line. So there is no point in hoping to overtake in a bend, you can't accelerate well there, and not just because the distance to cover is longer, but because you have to fight against centrifugal force. If we want to translate this into a formula, for a similar effort

$$v_{bend}^2 + \frac{\tau^2 v_{bend}^4}{R^2} = v_{straight}^2,$$

where τ is the friction coefficient per unit of mass and is of order 1. In the first lane, $R = 36.5$ metres, so compared with a speed of 10 m/s in a straight line, we go down to 9.66 m/s in a bend, maintaining the same effort. The tighter the bend (the smaller R), the greater the deceleration. The greater the speed, the greater the effect of centrifugal force. For example, the force is greater for horses running at 20 m/s in a 100 m radius bend than for a human at 10 m/s in a 36.5 metre bend.

This raises the question of why long races continue to be concentrated on lane 1, which is the most unfavourable for performance.

CHAPTER 2

THROWING

WHY ARE ALL JUMPS AND THROWS BASED ON THE SAME EQUATION?

(Painting by Godfrey Kneller)

Thanks to a falling apple, Isaac Newton formulated his laws of gravity in the 17th century. Objects as small as an apple or as large as a planet are subject to the same laws. And these laws also help us understand the trajectories and movements of sports, balls, ski jumpers, high jumpers, skaters, cyclists, and so on. The equations of motion can be written from Newton's second law (see box **Newton's laws**). Firstly, we need to determine the forces or actions exerted on the object, which then enables us to calculate the position and velocity of a body at any given moment as a function of its mass, initial position and velocity, and the initial effects it has been given.

It is therefore essential to identify all the forces involved in the movement of an object. The weight, which is always present,

© The Author(s), under exclusive license to Springer Nature Switzerland AG 2024

A. Aftalion, *Be a Champion*, Copernicus Books,
https://doi.org/10.1007/978-3-031-54082-0_2

Forces

A force is a cause capable of modifying the movement, trajectory or speed of an object, or deforming it. The intensity of a force is measured in newton, named after the inventor of the concept. A force is represented by an arrow (vector) indicating its direction, orientation, and intensity. The length of the arrow is used to analyze the effect on the trajectory. Forces can be either contact forces or forces acting at a distance, such as weight. In the former case, they are represented as acting at the point of contact, in the latter, they are represented as acting at the system's center of gravity (see next box). In the field of sports, we find:

- **the weight**: this results from the attraction exerted by the Earth. It is directed vertically downwards and acts on the whole body, but is represented as acting at the center of gravity.
- **ground reaction**: the force exerted by the ground or support on an object, directed upwards perpendicular to the surface.
- **friction:** any force resisting the sliding movement between two bodies, such as a bicycle wheel on the road. It is always opposed to the direction of motion.
- **tension**: the force exerted by a wire or rope.
- **fluid forces in air or water**:
  - lift, the upward force exerted by the buoyant force of Archimedes in water;
  - drag, a force resisting movement in air or water;
  - Magnus force, experienced by a rotating object in a fluid, which modifies the orientation of the trajectory.

By controlling forces, we control movement and performance.

results from the Earth's gravitational pull and is directed towards the ground. It is equal to mass multiplied by the acceleration of gravity g, which is approximately 9.81 m/s^2. The exact value of g depends on where you are on Earth (altitude, for example). The mass, on the other hand, is the same everywhere. For projectiles that are thrown, air resistance, also known as drag force, can be significant. It is proportional to the surface area exposed to the wind and to the speed of the projectile. Another, more complex force also comes into play, the Magnus force. If a ball is thrown with a rotational effect, it changes its trajectory into a spiral. According to Newton's second law, under the combined influence of these forces, the object speed and position change. Depending on the relationship between the forces (weight, air resistance and Magnus force), the type of trajectory is different.

Newton's laws

Isaac Newton established three very simple laws:

- The first law states that in the absence of forces, the speed of a body is constant. In particular, a body at rest remains at rest if no force is exerted on it, which seems natural enough. It can only start moving if a force is exerted on it. For example, if a puck has an initial speed on perfect ice, it retains this speed indefinitely. In real life, however, ice is never perfect; it creates friction which slows the puck down. This brings us to the second law.
- The variation in speed is equal to the sum of the forces exerted on an object. If forces do not compensate, they cause a variation in speed. In technical terms, this speed variation is called acceleration. In physics, deceleration is a negative acceleration. This law makes it possible to quantify and calculate variations in speed as a function of the applied forces.
- The last law is that of action and reaction. If a body A exerts a force on a body B, for example, a footballer kicking a ball, or a pole-vaulter leaning his pole on the ground, then body B exerts an opposite force of the same intensity on body A. The ball exerts a force on the player's foot, and the pole on the pole-vaulter. The pole allows the pole-vaulter to rise, but for soccer, the ratio of their two masses means that the footballer doesn't fall after kicking the ball.

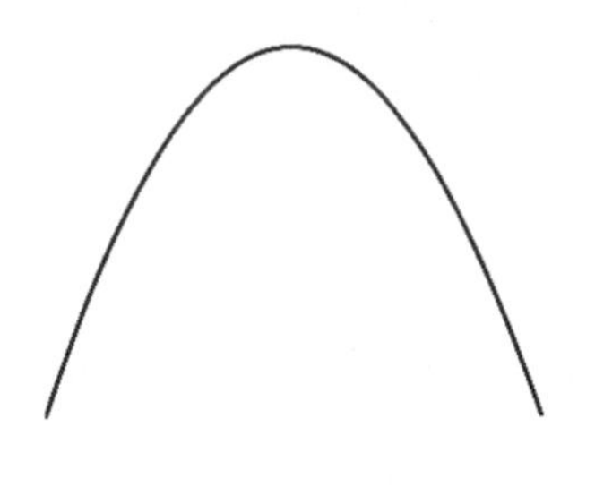

The simplest trajectory for an object thrown and subject only to its own weight is the parabola. It is symmetrical with respect to the point of maximum height. This is the trajectory of a basketball, the centre of gravity of a long jumper, a high jumper, a diver, a gymnast.

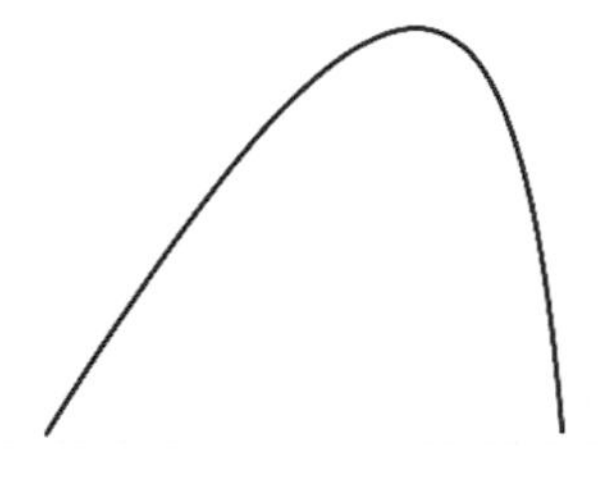

When the air slows down the movement, the trajectory rises almost in a straight line, then falls back down, almost vertically, with very little horizontal velocity. This is the trajectory of a badminton shuttlecock or a golf ball.

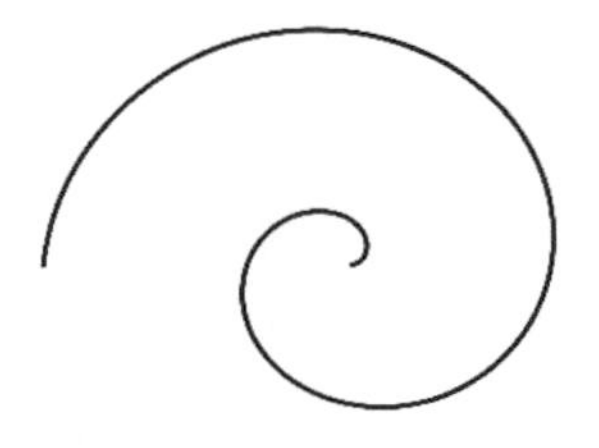

When the ball is rotated on itself and the main force is the Magnus force, the trajectory is spiral. This is the case for some soccer shots.

The general form of the equation governing movement in sports is therefore universal. It depends on the object only in terms of its mass, the position of its centre of gravity and the initial effects it is given.

WHY DO ALL BODIES FALL AT THE SAME SPEED?

Throws are used in a wide range of sports, including shot put, hammer, javelin, soccer, tennis, badminton shuttlecock, and the human body itself in the long jump, high jump, ski jump, parachute jump or basketball jump. The question of the optimal throw is linked either to the optimal throwing distance or to the minimum time spent in the air.

Here, we will analyze the case where the trajectory is mainly due to the effect of weight, and we will focus on other cases, with effects on balls, in the next chapter.

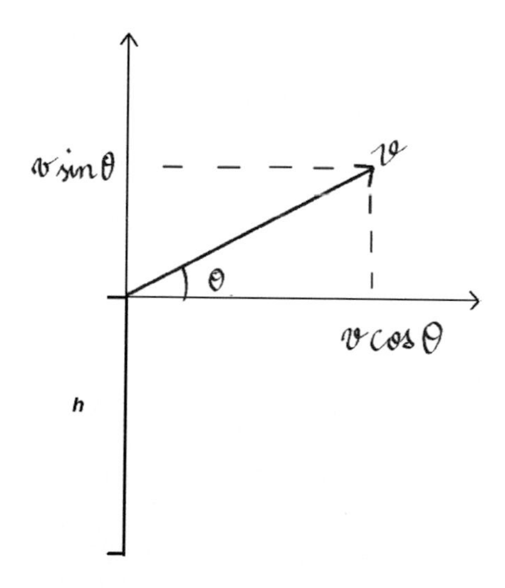

Initial speed of an object thrown from a height h. The speed v has an angle θ with the horizontal axis. The sine and cosine of θ allows us to compute the abscissa and ordinate of v; they are numbers between –1 and 1.

The simplest case is that of an object launched from a height h above the ground, without effect, without wind, with just an initial velocity v that makes an angle θ with the horizontal (see figure above).

According to Newton's laws, since there is no horizontal force (the weight is directed downwards), the horizontal velocity is constant. On the other hand, the vertical component of velocity is subject to the acceleration of gravity and its variation is proportional to time (according to the formula $v_y(t) = v_y^0 - gt = v\,sin\,\theta - gt$).

This component is initially positive as the object rises, then cancels out at the top of the parabola (at time $t = \frac{v \sin\theta}{g}$). The vertical velocity then changes sign, meaning that the object is falling down. Positions x and y (in the horizontal and vertical directions) can be calculated as a function of time t

$$x = v\cos\theta\ t, \qquad y = h + v\sin\ \theta\ t - \frac{gt^2}{2}$$

or replace t thanks to the 1st equation by $x/v\cos\theta$ to express y as a function of x in the 2nd equation :

$$y = h + x\tan\theta\ - \frac{gx^2}{2(v\cos\theta)^2}.$$

This is the equation of a parabola.

Let us go back to the time when the top of the parabola is reached, that is the maximal height (at time $t = \frac{v \sin \theta}{g}$), then

$$y_{max} - h = \frac{(v \sin \theta)^2}{2g}.$$

This maximal height does not depend on mass but only on the initial vertical speed ($v \, sin \, \theta$). This is why in the high jump, in order to jump high, it is crucial to transform the horizontal force created by the initial running into vertical velocity.

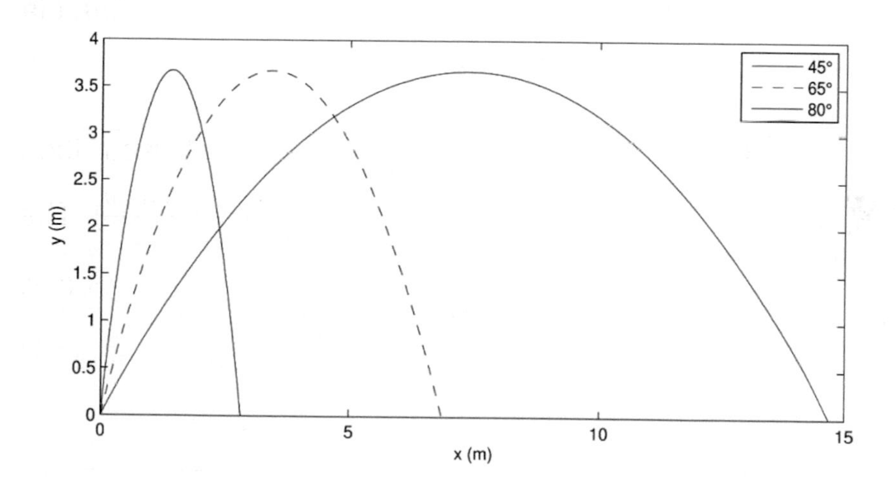

Trajectories with same initial vertical speed ($v \sin \theta$ is constant) but θ varies. The maximal height and time of flight are the same but the range changes.

One can also compute the time of flight (which is equal to the time where $y = 0$) and the range (value of x at this specific time). Let us start in a simple case where $\theta = 0$, so that the initial speed is only horizontal. Recall that the square root of

x, written $\sqrt{x}$ is the number such that $\sqrt{x} \times \sqrt{x} = x$. Then the time of flight is equal to

$$T_{flight} = \sqrt{\frac{2h}{g}}.$$

It is therefore independent of initial velocity and mass! Whether we let the object fall ($v = 0$) or send it with only horizontal velocity ($\theta = 0$), it will take the same time to reach the ground. However, it will not reach the same position. In the figure, we keep the vertical speed constant, so $v \sin \theta$ remains constant, but v and θ vary. The time of flight is the same, so is the maximum height, but the range changes.

Whether the object is heavy or light, it will take the same time to reach the ground. Note that this is not true for very light objects, such as a feather, because then the effect of air on movement is not negligible, but it is for a ball, a weight or a human being. The range, on the other hand, obviously depends on the initial velocity and is $p = T_{flight}\, v$.

WHY IS THE BEST ANGLE IN THE SHOT PUT 42 DEGREES?

When you throw an object from the ground with an initial velocity, and when you don't need to take air resistance into account, the optimum angle of inclination to send it as far as possible is 45 degrees, exactly halfway to the vertical.

A shot, a discus or a javelin are not thrown at ground level, but higher up, since the athlete's arm is about 2 m from the ground. As a result, the angle is reduced slightly to 42 degrees.

If we send an object of height *h* above the ground without effect, without wind, with just an initial velocity v that has an angle θ with the horizontal, we can calculate the positions x and y as a function of time t, as we have just seen:

$$x = v \cos\theta \; t, \qquad y = h + v \sin\,\theta \; t - \frac{gt^2}{2}.$$

Let us start with a simple case, where $h = 0$: the object is sent from the ground. Then the time of flight is the time when $y = 0$, that is

$$T_{flight} = \frac{2v \sin\,\theta}{g},$$

and the range, the value of x at this time

$$p = v \cos\theta \times T_{flight} = \frac{v^2 \sin\,2\theta}{g}$$

because $\sin\,2\theta = 2 \sin\,\theta \cos\theta$. Simple knowledge of trigonometry shows that the range is maximum when $\sin 2\theta$ is maximum, and thus is 1. This is the case for $2\theta = 90^\circ$, so $\theta = 45^\circ$. For a throw with no wind and no spin, the angle that produces the longest shot is 45°.

Note in the following figure that the angles whose sum is 90° (said to be complementary) produce the same range. On the other hand, the time of flight is very different. So a throw at 80° produces the same range as a throw at 10°, but the object stays in the air 5.7 times longer (tan 80°) and goes 32 times higher. In some sports, where it is strategic to send the ball into the air before running to catch it, it makes sense to throw at a higher angle (up and under kick) to have a long time spent in the air, and a short range.

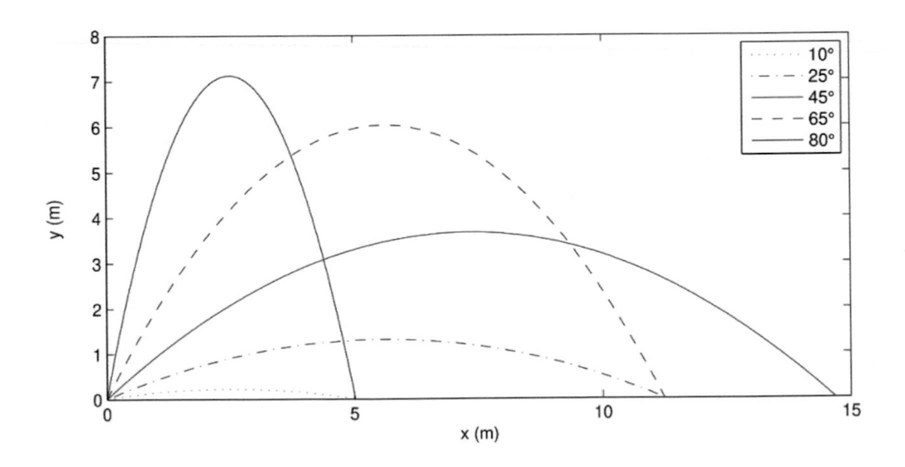

Trajectories with same initial speed but different angles. Complementary angles produce the same scope.

In many sports, the initial height of the projectile is not the same as its final height. For example, the shot putter holds it at a height h of around 2 m, whereas at the end of the trajectory the height is zero, since the object is on the ground. This has no impact on the analysis of forces and the calculation of velocities, but it does have an impact on the time spent in the air, since the shot falls back to a height lower than its initial height. The range can then be calculated as a function of the throwing height h, the velocity v, and the angle θ

$$\frac{v^2 \sin\theta \cos\theta + v \cos\theta \sqrt{(v \sin\theta)^2 + 2gh}}{g}.$$

Therefore, we can also calculate the optimum throwing angle as a function of h; the formula will be a little more complex, but this is precisely where mathematics (derivatives) become useful. We find, for example, that the optimum angle is 42° for a height of 2 m, an initial velocity of 13 m/s, and a range of 14.51 m. The advantage of having a mathematical formula like this is that it allows you to calculate the variations in range when all the parameters are modified by, say, 5%. If the launcher is not at the optimum angle but deviates by 5% (so 0.8°), then its range varies by 4 cm, which is very small. If he makes progress and is able to throw 5% higher, for example, at a height of 2m 10, he improves his throw by 11 cm. Finally, if he is able to throw 5% faster, i.e., at a speed of 13.7 m/s, he will improve his throw by around 1m 50. The most important parameter to work on is therefore initial speed rather than angle!

WHY DOES POSITION AFFECT LONG JUMP LENGTH?

At the 1968 Mexico City Olympic Games, Bob Beamon shattered the long jump record, improving on the previous year's record by 55 cm with a leap of 8.90 m on his very first attempt. While this is still the Olympic record, it has not been the world record since Mike Powell improved it, by just 5 cm, in 1991. To explain Bob Beamon's incredible performance, a favorable wind has been mentioned, or the high altitude of Mexico City, but the real explanation lies in his use of the centre of gravity.

There are sports, and long jump is one of them, where it is the body that undergoes a parabolic trajectory. Or more precisely, its centre of gravity. If you change the body's position, you move the centre of gravity, and this is what Bob Beamon has exploited!

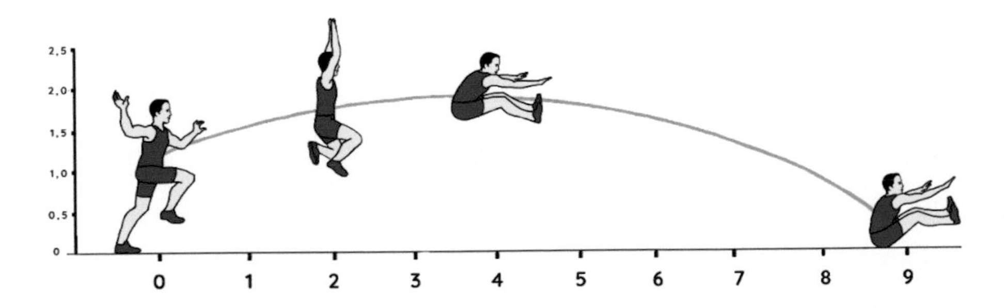

Long jump : the centre of gravity follows a parabolic trajectory. The jumper can modify the position of his centre of gravity by changing his body shape. Therefore, he can lower his centre of gravity and jump further.

At any given time, we can determine the "central" point which would replace the athlete and which has, on average, the same trajectory. This point is the centre of gravity or centre of mass, and depends on the athlete's shape and body position. It can be moved by modifying the latter, and this is crucial in the long jump. At impulse, Bob Beamon's centre of gravity is about 1 m 20 from the ground. He takes off with a velocity $v = 10.02$ m/s at an angle θ of 23 degrees. His centre of gravity follows a parabolic trajectory, so if he did nothing to move it and fell back into the same position, his jump would have a range of $\frac{v^2 \sin 2\theta}{g}$, or 7 m 36. However, by adjusting the height of his centre of gravity when he touches the ground, he can significantly increase the length of his jump. On landing, his centre of gravity is not 1 m 20 from the ground but 50 cm, a difference of 70 cm. The maximum range of his jump is therefore greater, and the above formula gives us a good estimate:

$$\frac{v^2 \sin\theta \cos\theta + v \cos\theta \sqrt{(v \sin\theta)^2 + 2gh}}{g} = 8.75\,m.$$

Centre of gravity

1. Where is it?

The centre of gravity or centre of mass is an imaginary point that allows us to understand the effect of forces on a complicated system, taking into account the distribution of masses. For a ball, it's at the centre. For a dumbbell made up of two balls of the same mass, it's in the middle of the segment between the two balls. This is the easiest place to grip the dumbbell in a balanced way. But if the two masses are different, for example if one is twice as heavy as the other, the ideal place to grip the dumbbell will be two-thirds of the way towards the heavier mass.

If we complicate things a little and imagine three masses, the centre of gravity is that of the triangle, located at the intersection of the straight lines joining one vertex to the middle of the opposite side.

To transpose this to the human body, we need to imagine that we can divide the body into several zones of fixed mass, each with a centre of gravity, such as the arms, legs, torso, head ... Depending on the relative position of all these elements, we find the point corresponding to the centre of symmetry of the set of points found, each weighted by their mass. We would expect the centre of gravity of a very lean person, whose mass is uniformly distributed, to be towards the middle of the body, at the level of the navel. If the athlete changes position, the centre of gravity shifts.

Sometimes, this fictitious point can be outside the body as is the case of a gymnast leaning on the arms or n a bridge, for example, or of a high jumper in Fosbury, whose centre of gravity is located below him, and often under the bar he is clearing.

If you raise your arms or bend your legs, your centre of gravity rises. Lifting the leg backwards (as when a gymnast performs an arabesque) shifts the centre of gravity backwards. To keep it more or less in its original position, the opposite arm must be raised.

Depending on the gymnast's position, the position of her centre of gravity varies. Her centre of gravity is indicated by a cross. It can be outside the body.

A basketball player who bends his legs and raises his arms. His centre of gravity rises by nearly 10 cm.

The centre of gravity thus shifts according to the body's movements and position. One of the strengths of the best athletes is their ability to control the precise position of their centre of gravity to optimize their performance.

2. Balance

For a person to be in equilibrium, his centre of gravity must be above his "support polygon", which is the figure formed by connecting the points of support on the ground. For example, for a person whose legs are spread, the support polygon is the figure formed around his two feet, and the centre of gravity must be vertical, above this figure. For a gymnast leaning on her hands, the support polygon is the figure formed by the hands, and the centre of gravity must be vertical, above this figure.

A gymnast who has changed the position of her centre of gravity so that it is no longer above the bar, for example by stretching her arms, undergoes a rotational movement and falls. Similarly, a diver with only the toes on the diving board is in a falling situation, as his or her centre of gravity is no longer vertical above the feet.

Gymnast in balance: her centre of gravity is aligned with the vertical of the weight. If the centre of gravity is not aligned with the vertical of the weight, the gymnast falls in rotation around the bar.

How does the long jump athlete lower his centre of gravity on landing? He extends his legs and arms horizontally, far out in front of him, and reduces his vertical "thickness".

One might ask why he doesn't choose a jump angle greater than 23 degrees to get closer to the optimal 43 degrees. In fact, as we have seen, what counts above all is speed. The horizontal speed of long jumpers is close to that of sprinters, i.e., of the order of 10 m/s. On the other hand, it is impossible for them to produce the same speed vertically: the maximum vertical velocities of basketball players, for example, are of the order of 4 m/s. For horizontal and vertical speeds to be identical, the horizontal speed would have to be considerably reduced, and performance would suffer. It is not possible to optimize all parameters, so it is important to choose those that have the greatest influence; in the case of a jump or throw, the speed rather than the angle of inclination.

WHY DO BASKETBALL PLAYERS SEEM TO HANG IN THE AIR?

The jump shot is one of the most important shots in basketball. It is a basic shooting technique in which a player throws the ball towards the basket from a **straight vertical jump**. Most athletes give the illusion of floating at the top of their jump, standing still, as if suspended in mid-air. This is why Michael Jordan is said to fly.

The basketball player's centre of gravity has a parabolic trajectory. The head remains at the same height during the jump. It is the centre of gravity that rises when the arms are raised and the legs bent, giving the effect of suspension in the air.

Once again, it is an effect of the centre of gravity! The athlete's centre of gravity always follows a parabolic path through the air, constantly changing position. But let us imagine that the basketball player, when he is at the top of his trajectory, manages to raise his centre of gravity, bringing it closer to his head. How does he do this? He bends his legs and raises his arms. So, for a moment, the centre of gravity has a parabolic trajectory, but as it moves through the body, the head remains at a constant height.

As the audience's attention is focused on the athlete's head, and the head remains at the same height for some time, this produces the effect of suspension in the air.

It is the same for ballet dancers performing a grand jete. At the top of their trajectory, they raise their arms, which raises their centre of gravity and allows their head to remain at the same height for a few moments.

WHY DO YOU JUMP HIGHER IN FOSBURY THAN IN SCISSORS?

In 1968, at the Olympic Games in Mexico City, while Bob Beamon was breaking the long jump record using the properties of the centre of gravity, the high jump also underwent a revolution. Dick Fosbury broke world record with a jump at 2 m 24, by introducing a new technique. He rose with his back to the bar, passing his head first, then his loins and buttocks. His body was then curved, with his head lower than the bar and his legs dangling. Once approved by the judges, this new technique became known as the Fosbury flop or back roll.

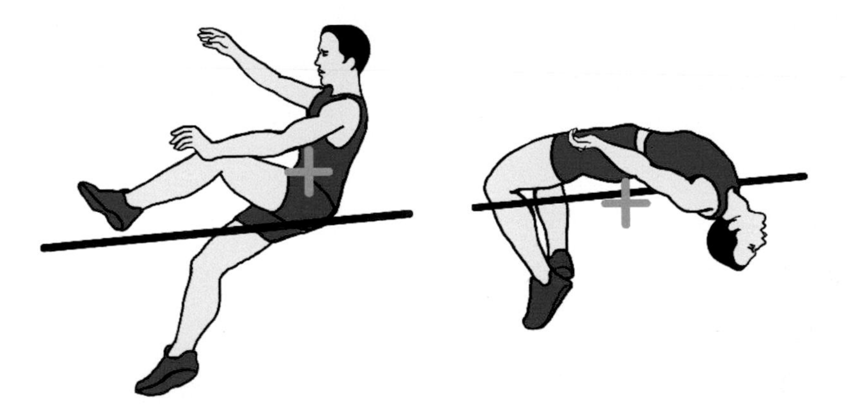

Left scissor jump: centre of gravity is above the bar. Fosbury jump to the right, centre of gravity below the bar.

In the high jump, the aim is to raise the body as high as possible. It is the body that must rise above the bar, not necessarily the centre of gravity. The back roll enables you to jump with your centre of gravity lower than the bar. In fact, if properly executed, by the time your hips get above the bar, your head and legs are below the bar itself, like the athlete's centre of gravity. In the end, a jumper could clear 2 m 45 (Cuban Javier Sotomayor's world record since 1993), for example, without his centre of gravity ever reaching that height!

Consider a turn-of-the-century athlete performing a scissor jump. His centre of gravity is above his navel. Therefore, in order to clear the bar, he has to raise his centre of gravity much higher than the bar. In the case of a Fosbury jump, the centre of gravity is under the bar around which the body wraps, so you don't have to raise it as high as a scissors jump to clear the same height!

In order to jump as high as possible, the athlete must maximize his vertical jump speed take-off. The optimal technique is to lean back a little and plant the supporting leg forward, pushing very hard on the ground. The jumper must reach a high speed before take-off, which enables him to maximize the vertical force exerted by the leg, and contributes to maximizing vertical speed. The supporting foot acts as a pivot, transforming part of the horizontal momentum into a vertical take-off speed. The value of this vertical take-off speed is of the order of 4 to 5 m/s, when a runner can reach 10 m/s. We are therefore far from being able to transform all the speed of the race into vertical propulsion!

WHY DOES A POLE ALLOW YOU TO RISE TO 6M?

Pole vaulters also have to transform their horizontal run into vertical propulsion, with the pole this time being the crucial element.

After the era of Soviet champion Serguei Bubka, who set 17 successive world records (from 5.85 m in 1984 to 6.14 m in 1994), France's Renaud Lavillenie cleared 6.16 m on his first attempt on February 15, 2014, and Sweden's Armand Duplantis supplanted him in 2022 with 6 m 21 and in 2023 with 6 m 23. How did they manage it?

Pole vaulting has several phases:

- The run-up phase. The athlete runs to reach the highest possible speed. Champions reach running speeds similar to those of sprinters. The weight of the pole, between 3 and 5 kilos, slows them down. To limit this effect, the pole vaulter holds it upwards. In this way, the so-called lever arm, i.e., the downward torque exerted by the weight, is reduced (see Chapter 5). You might think that the further you hold the pole from the end, the more you reduce this lever arm, but in fact you mustn't move too far from the end to succeed in the next phase.
- The take-off phase. The pole vaulter locks the pole in the pit.

- The pole bending phase and the pole straightening phase. The pole flexes, straightens and takes the athlete upwards. There is a double pendulum movement. The pole rotates around its fulcrum, and the pole vaulter rotates around the pole, moving from head-up to head-down, in a circular motion around his hands. The elasticity of the pole is essential.
- The jump. Pole vaulters use their arms to propel themselves up the pole, gaining around 20 cm in the process. To maximize jump height, athletes wrap themselves around the bar (as in the high jump). Their centre of gravity can then be located under the bar, while allowing their body to pass over it. The bar can therefore be placed higher than the maximum height reached by the jumper's centre of gravity.

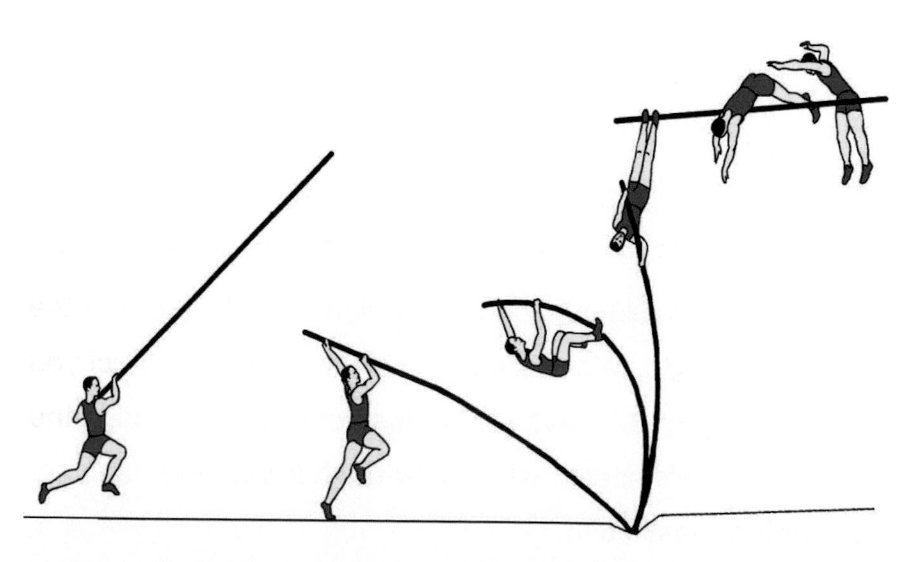

Pole vault phases. The run down, take-off, pole bending, pole straightening and jump.

It is important for the pole material to be elastic, so that energy transfer can be quick and effective. To achieve this, poles are made from composite materials that are both strong and flexible. It would be conceivable to build longer poles, but they would bend under their own weight and have difficulty straightening out completely. If the pole is too rigid, its impact on the ground results in a loss of energy for the athlete, who is propelled backwards. If it is too flexible, energy is not released quickly enough and the pole is unable to propel the athlete.

We know that Sergei Bubka ran at around 9.9 m/s. Assuming that his centre of gravity is 1 m high, that he gains 20 cm thanks to the thrust of his arms, this means that the pole makes him gain 6.14 - 0.2 - 1 = 4.94 m. Physics tells us that, to lift a mass m from a height h, you need a potential energy equal to $E = m \times g \times h$, where $g = 9.81\ \mathrm{N/kg}$.

If we imagine that the pole does its job perfectly and that all horizontal kinetic energy (due to running speed) is transformed into potential energy, then $\frac{1}{2} m\, v^2 = mgh$. The same height would be found if v was exactly the *vertical* propulsion speed. We therefore find that to rise 4.94 m, we need a speed of 9.85 m/s (regardless of mass!). This is very close to Bubka's speed, which means that the pole has very good energy restitution.

Let us remember that during the run-up, the vaulter increases his speed and accumulates kinetic energy, which is transformed during the jump into energy for bending the pole, and then restored in the form of potential energy, i.e., for rising. It is conceivable that pole vaulters could gain a few more centimeters by propelling themselves harder over the pole with their arms.

WHY DO YOU JUMP TO SERVE IN VOLLEYBALL?

One way to optimize a volleyball serve is to minimize the time the ball spends in the air, thereby reducing the reaction time of the opposing team and making it more difficult to return the shot. But how to minimize the time the ball spends in the air after the serve? There are three important points:

1) let the ball be just above the net, as low as possible,
2) serve so that the ball lands at the back of the court,
3) serve with a jump to get the ball off as high as possible.

In volleyball, as a good approximation, we can assume that the weight is the dominant force and the trajectory of the ball is a parabola. As we saw earlier, the higher you start from (i.e., by jumping), the longer the range for a given speed and angle, so the better the shot in volleyball. The ball also needs to pass over the net and not fall back too quickly, which requires an almost horizontal shot. A bell-shaped shot would stay in the air longer and therefore would be easier to return. The next figure shows the trajectory of the ball from the point of service to the point of landing as a function of the height of the service h_o, the height of the net H and the maximum height reached by the ball h_{max}.

Point A is the service location, with coordinates $(0, h_o)$, point B is just above the net, coordinates (L_a, H) and point C is where the ball lands, coordinates $(L_a + L_b, 0)$. Note that L_a is

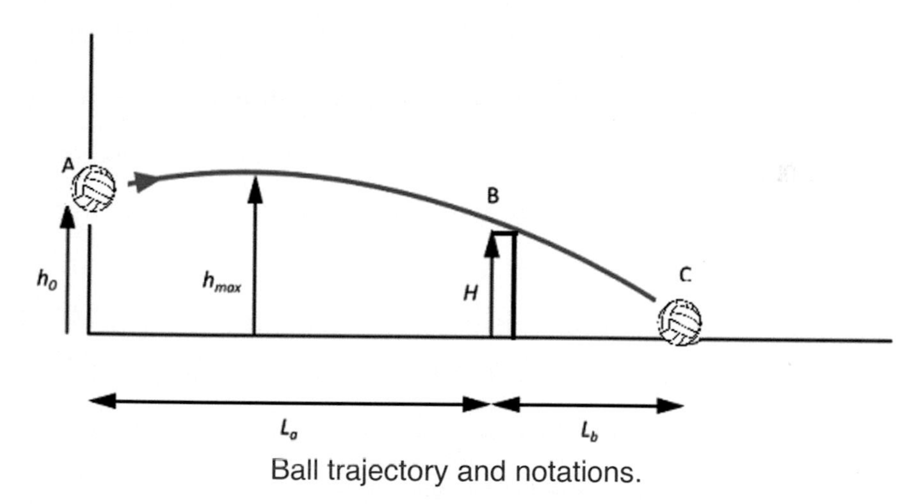

Ball trajectory and notations.

the distance from the service location to the net, in the direction along which it is served, and L_b is the distance from the net to where the ball lands on the other side of the court, in the direction of the service. We choose $L_a = 9$ m (court size) and h_o, L_b are to be optimized, as well as v the throwing speed and θ the throwing angle.

For simplicity, we will assume that air resistance and aerodynamic effects acting on the ball can be ignored. From the equations of motion, we have, if x and y denote the ball's position at any point along its trajectory,

$$y = h_o + x \tan\theta - \frac{gx^2}{2(v\cos\theta)^2}.$$

The coordinates of points *B* and *C* can be used to express the initial velocity v and the initial angle θ as a function of L_b. The time the ball spends in the air (which we want to minimize), to get from *A* to *C*, is given by $\frac{L_a+L_b}{v\cos\theta}$. As θ and v depend on L_b, this formula cannot tell us how to minimize time directly. A few calculations show that we need to minimize the maximum height to which the ball rises so that it spends less time in the air. However, for the ball to land close to the net (with a small L_b), we need a high arc. This means that the ball is suspended in the air for a longer period of time. It can be shown that we need to make L_b as large as possible (serve so that the ball lands at the back of the court with a large diagonal). Serving while jumping makes h_o as wide as possible and allows the ball to start its downward trajectory earlier (since h_{max} is reached earlier). To get an idea of this time, let us say we have L_b = 9 m, h_o = 3.0 m and H = 2.4 m, then t = 0.86 seconds. This is very quick.

Finally, giving the ball a spin on itself can save a tenth of a second. We will look at these effects in the next chapter.

CHAPTER 3

BALLS

WHY CAN BALLS SEEM TO DEFY GRAVITY?

If we ignore the effect of air, the movement of the centre of mass is the same for all objects and all athletes: it is a parabola. But many sports rely on the effect of air which dominates over gravity when objects are light, moving fast or spinning quickly. Without air, there would be no slice in golf or cut balls in table tennis. These lifting, curving and slowing forces are complicated. We will analyze mainly the slowing force on a fast object (which depends on the shape, texture and velocity) and the curving force on a rotating object, which can bend the trajectory to the left or right, or up or down.

When only gravity is involved, the equations of motion can be solved analytically with formulas. Where air is involved, however, the change in velocity (acceleration) depends on the velocity itself. The equation of motion is then non-linear, cannot be solved analytically, and requires a computer to calculate its solutions numerically. We can get an idea of the phenomena involved by making approximations and assume, as is classic in physics, that one force dominates over the others.

© The Author(s), under exclusive license to Springer Nature Switzerland AG 2024

A. Aftalion, *Be a Champion*, Copernicus Books,
https://doi.org/10.1007/978-3-031-54082-0_3

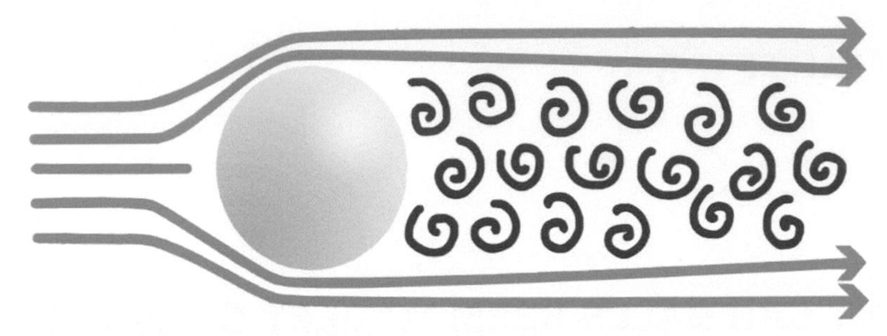

Ball moving to the left. The air hits the ball, threads of air go around it and vortices form in its wake.

A very important force in the air is drag, which is opposed to movement and depends on speed. To understand what this is, we need to imagine that the air is made up of lots of tiny particles that collide with the ball, as if the ball was thrown onto a carpet covered with tiny marbles. This slows the movement down. This braking force obviously depends on the number of collisions, i.e., the number of atoms in the air, the surface area of the ball and its speed. Indeed, the faster it goes, the more collisions there are per unit of time, and therefore the more it is slowed down. A formula has been established for the force of drag, as follows:

$$\frac{1}{2}c_d A v^2,$$

where A is the surface area of the ball, v its velocity and c_d is the drag coefficient, which is usually measured experimentally and depends on the texture and type of the ball. For a smooth ball, it is around 0.5 but can be reduced depending on texture and speed, as we will see below.

The description of collision drag is correct at low speed, i.e., under 20 mph. But as speed increases, the description needs to be refined. Air hits the ball at the front, air streams around it and vortices form in its wake. It is these three phenomena that make up drag.

Drag coefficient

The drag coefficient is the numerical coefficient that appears in the drag force calculation

$$\frac{1}{2}c_d A v^2,$$

where A is the surface area of the ball and v is its velocity. When a ball is moving slowly, the drag coefficient c_d is of the order of 0.5, and the flow is then said to be laminar. There is a layer of air around the ball, moving forward with it. This layer of air separates at the ball's equator, as shown in the figure, leaving a wake behind the ball, where a vacuum is formed that displaces the air. Pressure is therefore lower behind the ball, and greater in front of it. This pressure difference slows the ball down. The size of the wake at the rear is linked to the drag coefficient. A wide wake corresponds to a high drag coefficient. The faster the ball moves, the more turbulent the air layer becomes, allowing air to flow more easily so that the wake is narrower. The drag coefficient of a very fast ball becomes smaller. It can reach values of the order of 0.2 for a smooth ball at 150 mph. But while the drag coefficient is smaller at high speed (divided by 2 or 3), the drag itself is greater, since it depends on the square of the speed.

Smooth balls have a greater drag than uneven balls. Indeed, irregularities on a ball create small vortices that form a "skin" around the ball, improving air circulation and reducing the drag coefficient, which can then fall to 0.1 for speed around 60 mph. Seams on a cricket ball or baseball, for example, reduce drag. On the other hand, for a tennis ball, because of the nap (fuzzy cover of the ball), drag is greater (coefficient of the order of 0.65, but 0.5 for a worn ball). The nap does serve a purpose, however; it gives the racket better grip when struck.

As to the elongated shape of a football, it provides much lower drag than spherical ball presenting the same area to the wind. This is due to much lower values of c_d, around 0.2 for velocities above 10 mph.

At low speed, the drag does not influence the trajectory, but as the speed increases, it can become comparable to the weight, up to a speed limit sometimes reached by tennis, cricket, baseball or badminton balls, where drag and weight offset each other.

Similar trajectories can be observed in golf and badminton. The ball starts at a wide angle and accelerates to a speed limit towards the top, then falls back at almost constant speed as gravity and air resistance compensate.

Another force is exerted on a ball rotating on itself, the Magnus force, which is perpendicular to the speed of motion. If the ball spins clockwise, for example, the bottom of the ball has a greater velocity relative to the air than the top of the ball, since the velocity relative to the air is the opposite of the ball speed of travel, to which we add the speed of rotation. The wake is therefore deflected downwards (see the figure below). Bernoulli's law relates speed to pressure, with high-speed zones having lower pressure than low-speed zones. As we can see, the top of the ball feels a higher pressure than the bottom, so a force is exerted downwards, perpendicular to the initial speed and deflects the trajectory.

If the ball is rotated, as the figure illustrates (this is called brushing the ball), it curves its trajectory to the right, or to the left if rotated in the opposite direction. If the ball is brushed from top to bottom (called a "cut ball"), it undergoes an upward force, its trajectory lengthens and it has lower bounce. If, on the other hand, the ball is brushed from bottom to top (in the case of a lifted ball), a downward force is exerted, the trajectory is shorter and the ball has a higher bounce.

The formula that gives the intensity of the Magnus force is similar to that for drag:

$$\frac{1}{2}c_M\rho_{air}Av\omega,$$

where A is the surface area of the ball, ρ_{air} is the density of the air, v is the speed of the ball, ω is the speed of rotation of the ball on itself, and c_M is the Magnus coefficient. Air density decreases with altitude, so the Magnus effect is less pronounced at higher altitude.

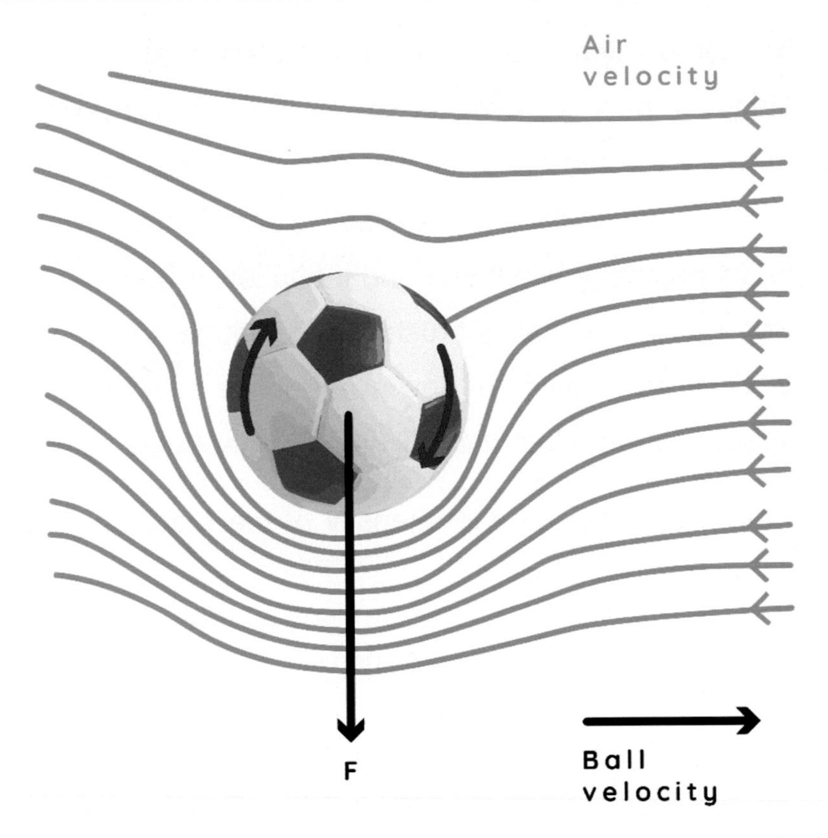

Illustration of the Magnus effect. The ball moves to the right (top view). As it turns on itself, the top of the ball feels a higher pressure than the bottom, so a downward force is exerted.

WHY DO GOLF BALLS HAVE DIMPLES?

The surface of a golf ball is not smooth, but made up of cells between 250 and 500 depending on the ball. Not only do no two balls have the same number of dimples, but the size, depth and shape of these dimples also vary from one ball to another. Understanding the optimal geometry and distribution of cells on a ball is a current research topic. This involves very time-consuming calculations, and the solution is not yet known.

Dimples on a golf ball.

These cells have a significant effect on the ball trajectory, reducing the frictional force (drag) that slows down movement. Drag varies greatly according to the distribution of dimples. Balls without dimples therefore travel a shorter distance than those with dimples, but if the dimples are too deep, they slow the ball down, and balls with the shallowest dimples are the ones that travel the furthest. More precisely, dimples must be shallow and the angle between the dimple and the surface of the ball must be marked, i.e., the dimple forms a break in the surface. Irregularities help to reduce drag. It has been calculated that, ideally, there should be a variation of the order of 40% in the slope in the transition zone where the honeycomb is dug into the ball, but with a very shallow depth. There are balls with circular or hexagonal honeycombs, distributed over repeating triangles or in a circular fashion, some with well-studied asymmetries, but no mathematical optimization theory has been able to select the best honeycomb distribution. In general, however, molds are made for two hemispheres, so there are never any cells on the equator!

At the beginning of the 20th century, the Englishman William Taylor realized that used balls went further than new ones, and that irregularities, or defects, improved range. But why does air resistance decrease with dimples?

When a ball moves, the air forms a thin layer around it, which moves with it. The ball cells generate small vortices all around it, allowing the air to "cling" to the ball, forming a kind of skin. It is as if the material on the ball's surface were air, so friction is much lower than with a smooth ball. Therefore, the wake behind a golf ball is smaller than behind a smooth ball.

Drag force is generally proportional to speed squared. In the case of a golf ball, the drag coefficient c_d (see the **Drag coefficient** box) is not constant but inversely proportional to speed. So, thanks to dimples, the drag force of a golf ball is proportional to speed (and not to the square of speed like a smooth ball), which for speeds of the order of 125 mph, considerably reduces drag compared to a ball without dimples!

The fact that drag is proportional to velocity rather than its square modifies trajectories in relation to the parabola. The ball rises almost linearly and falls almost vertically.

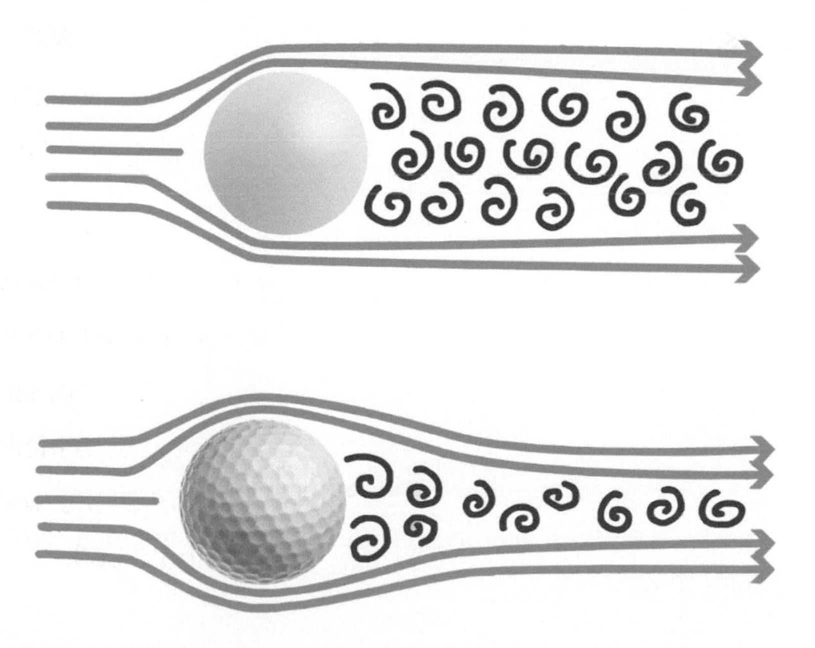

Dimples on the surface of a golf ball lead to a smaller wake (bottom) and lower drag than a smooth ball travelling to the left at the same speed (top).

We can write the equations of motion of the golf ball and solve them as a function of the shot angle θ_0 and initial velocity v and plot the trajectory for a shot angle θ_0 between 10 and 50° as shown in the next figure. The range, i.e., the maximum distance reached by the ball, is maximum for an angle of 26°. In high-angle trajectories, the ball starts, rises, slows down towards the apex, then falls back at almost constant speed (as weight and drag compensate). The trajectory is not at all symmetrical with the point of maximum altitude.

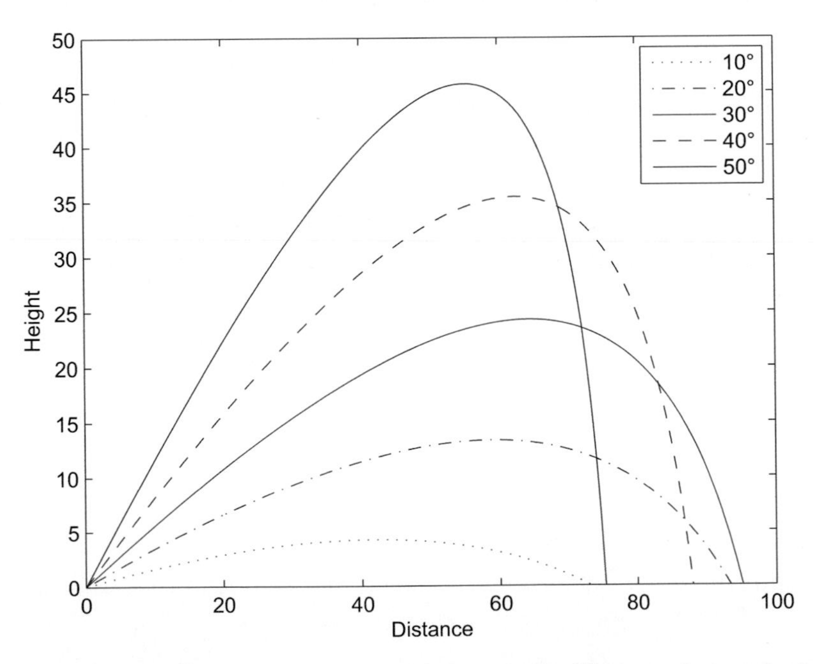

Trajectory of a golf ball according to the angle. The angle producing the longer range is 26°.

The angle θ_0 depends on the opening angle of the chosen club. In fact, experimentally, it is slightly less than the club angle. Golf club manufacturers generally state that maximum reach is achieved by woods with an optimum angle of 11°. To find this angle, which is less than 26° (above figure), we need to take lift into account. Indeed, the ball is often sent spinning on itself, creating a Magnus force that lifts it. This creates a pressure difference between the upper and lower surfaces. The rotation accelerates the air at the top and slows it down at the bottom. According to Bernoulli's principle, the pressure above the ball is therefore lower, creating a force that pushes the ball upwards. This force is expressed as

$$\frac{1}{2}C_l A\rho v,$$

where C_l is the lift coefficient, v is the velocity, ρ is the air density, and A is the cross-sectional area of the ball. A ball spinning on itself will therefore rise higher than a ball launched without effect at the same angle.

The equations cannot be solved explicitly in the general case, but the trajectories can be calculated with a computer. The two effects combined increase the length of the ball trajectory by around 30% compared to a smooth ball without dimples.

But for most golfers, the problem is not so much the exact angle of impact as hitting the ball in the right place at the right time!

WHY ARE SOCCER BALLS MADE OF HEXAGONS AND PENTAGONS?

The first soccer balls were made of six to eighteen leather faces held together by a lacing system. Inflated with air through a pig or sheep bladder, they were not perfectly spherical, which prevented them from rolling properly and could lead to faulty bounces. The first balls without valves or laces became official in 1952. From the 1960s onward, for the sake of TV visibility, balls were white, then for the 1970 World Cup, they were white and black. This black-and-white ball has been retained as the soccer symbol in many cases, for example for the ⚽ character representing soccer in emojis, or more generally in the Unicode UTF-32 (32-bit Universal Transformation Format) computer standard, coded U+26BD.

In its most widely represented form, and in order to make it as spherical as possible, soccer balls are ultimately made up of twenty white hexagons joined together by twelve black pentagons. But why is it designed this way? It is because these two polygons are the closest you can get to a sphere! Indeed, approximating a spherical shape with flat surfaces is a very complicated problem. While it is possible to make a solid with pentagons only, (but then it is a long way from a sphere and cannot roll very well), it is impossible to approximate a sphere with identical hexagons; you have to add pentagons. A soccer ball is therefore a 32-panel "polyhedron" which resembles a sphere as closely as possible. It is a truncated icosahedron. It has 32 faces (12 pentagons and 20 hexagons), 60 vertices (with 2 hexagons and 1 pentagon around each vertex) and 90 edges. Let us be more precise.

An icosahedron is an object with 20 facets (twenty = icosa in Greek): a crown of 10 triangles, with a "hat" composed of 5 triangles above and below (next figure on the left). It is said to be truncated because each of its 12 vertices has been cut to one-third of each of its edges (figure in the middle) so as to "round off" the ball finally obtained.

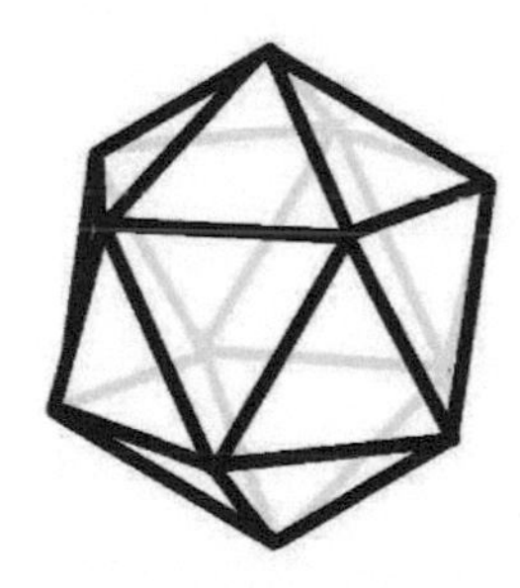

Truncated icosahedron: on the left, icosahedron, a polyhedron with 20 faces; in the middle, the 12 vertices are truncated at one-third of the edges; on the right, the figures obtained by truncating the vertices have been coloured black.

In the middle figure, the part removed (coloured black) reveals hexagons (grey). In the figure on the right, the figures colored in black have been truncated around the vertices, to form pentagons.

Once the ball is inflated, the edges and vertices disappear entirely, giving the ball an almost perfect circular shape.

Soccer ball made up of white hexagons and black pentagons.

The aim is to get as close as possible to a sphere so that the ball rolls smoothly, but the 32-panel shape has two other advantages. Firstly, it offers goalkeepers a better grip than a smooth sphere without roughness, which would slide around, and secondly, the irregularities produce more effect when the ball is kicked.

The shape of the balls continues to evolve with each World Cup. For the 2022 World Cup in Qatar, Adidas had created a new 20-panel ball called "Al Rihla", which means "the journey" in Arabic. It consists of eight rounded triangles and twelve panels in the shape of ice cream cones, inspired by the shape of the sails on certain boats.

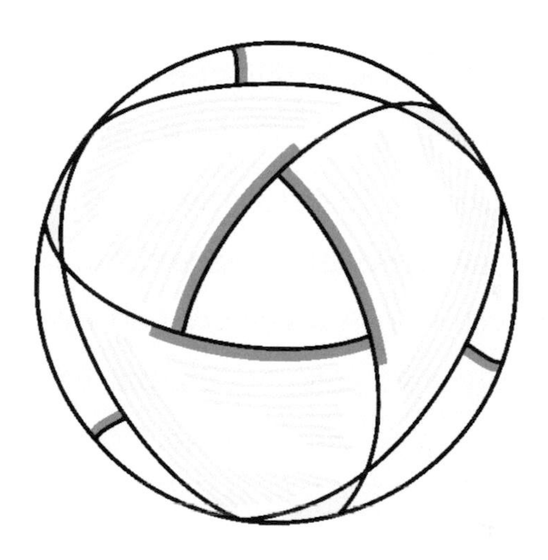

2022 World Cup Al Rihla ball.

The aim of the new ball is to make it go faster, and in particular to reduce drag at high speed. To achieve this, we have seen (see **Drag coefficient** box) that a slightly irregular surface is required. We need to find the right balance between the number of panels and their shape, the seams and the texture of the surface. This new ball has no seams and is heat-welded, but the joints are 2.5 mm wider and 0.5 mm deeper than previous balls, creating irregularities that limit drag. The coating (mostly polyurethane) and diamond-shaped embossing have also been designed to optimize speed. Research and technology progress with each new World Cup!

HOW DID BECKHAM SCORE FREE KICKS?

On October 6, 2001, England was trailing 2-1 to Greece and facing the prospect of a tricky play-off against Ukraine to reach the 2002 World Cup finals. In the 93rd minute, the captain and Manchester United favourite, David Beckham, stepped up to curl home a stunning 25-yard free-kick and sent England directly to the finals in South Korea and Japan. Greek players looked at the ball, whose trajectory had bypassed the wall. They thought they were safe, but the ball bent to the left and finally entered the goal, much to the dismay of the stunned goalkeeper and the BBC commentator.

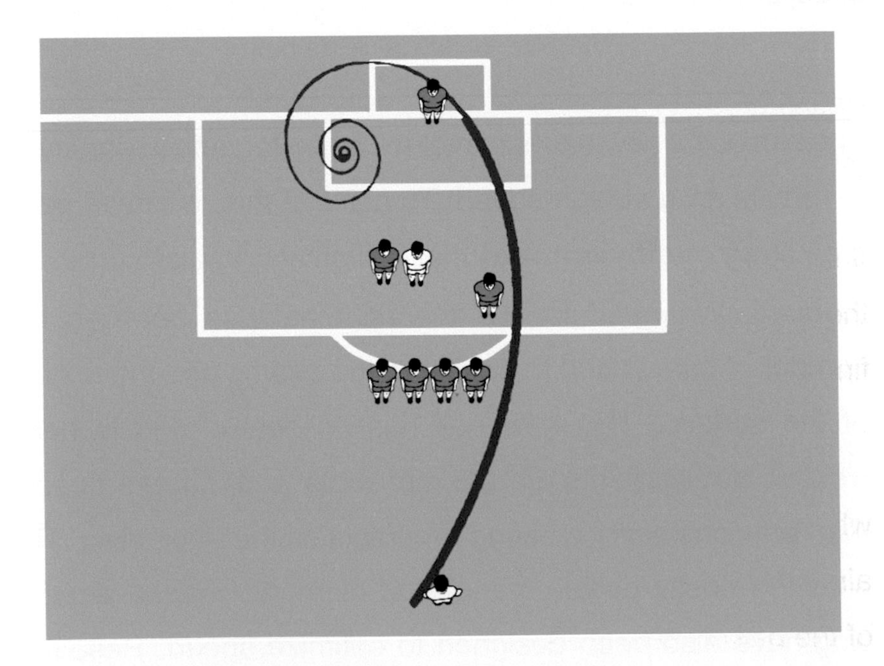

Free kick. Ball shot with Magnus effect, top view. The ball rotates on itself and follows a spiral trajectory. It bypasses the wall formed by the opponents, avoids the goal keeper and enters the goal. The thinner line is the end of the trajectory as if there were no net.

David Beckham was so good at taking free-kicks that the phrase ‘bend it like Beckham’ was popularised. But his famous free-kicks would not have been possible without the Magnus effect!

To achieve this, you need to throw a ball at a certain speed, spinning it on itself. If the ball is rotated counter-clockwise, its rotation accelerates the air on the left and slows it down on the right. By Bernoulli's principle, the pressure is therefore lower on the left side of the ball, creating a force that pushes the ball in this direction. If the ball is rotated in the opposite direction, the force is on the other side.

Beckham "brushed" the ball, for example by hitting it to the right of its axis, thus creating a rotation of the ball on itself.

If we imagine that the trajectory is almost circular, we can use the formula for Magnus force and centrifugal acceleration to find the radius of the circle

$$R = \frac{2m}{c_M \rho_{air} A \omega},$$

where A is the surface area of the ball, ρ_{air} the density of the air, v the speed of the ball, m its mass, ω the speed of rotation of the ball on itself and c_M the Magnus coefficient. Since drag must also be taken into account, the velocity decreases along the trajectory, as does the radius, creating a spiral trajectory.

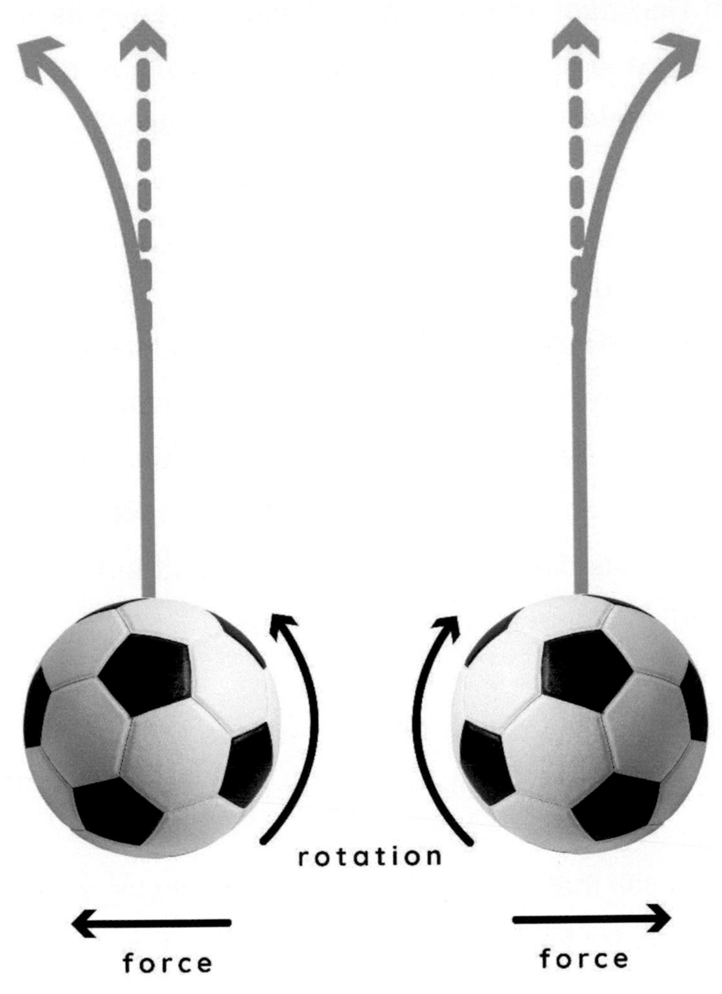

Overhead view. The ball is kicked upwards with a rotational effect. On the left, rotation to the left and trajectory deflected to the left. On the right, rotation to the right and trajectory deflected to the right.

Note that at higher altitudes, the density of the air decreases, which reduces the effect in question.

WHY DO BOWLERS SHINE ONLY ONE SIDE OF THE CRICKET BALL?

Cricketers shine a cricket ball because it helps in swinging the ball. Swing bowling is one of the oldest art forms in cricket.

A cricket ball with seams and a baseball with stiches.

If the ball is hurled in such a way that on one side of the ball the seam is more towards the front than on the other, the ball will experience a force on one side that causes the ball to accelerate that side. The side that has the shine provides little resistance to the wind, while the rough side resists the flow of the wind. The fast moving wind on the shiny side hits the raised seam and pushes the ball in the opposite direction. The result is a curve in the ball path called a swing.

To make sure the ball orientation does not change during its flight, the swing bowlers put a strong backspin about the ball's north pole (if the seams are considered the equator).

Over the course of the game, the hard leather surface of the ball becomes increasingly scuffed. The bowler recognizes that one side is more scuffed, and therefore works to preserve the asymmetry by shining the smooth side, rubbing it on his shirt, or even spitting on it. Indeed certain candies are supposed to produce a saliva quite well suited to polish the ball.

While pitchers also put a spin on a baseball to make it curve, the purpose of the spins are very different in cricket and baseball. In cricket, spin is used to keep the seams pointed at a constant angle and it is the asymmetry in the seams that then gives the swing. In baseball, it is the spin itself that causes the ball to curve thanks to a Magnus effect. Note that the seams in baseball can lead to an erratic effect. Indeed, if the baseball is spinning very slowly, the lateral force will point first to one side and then to the other side according to the seam direction making the ball essentially dance or flutter in the air. This is called a knuckleball and is very difficult to achieve. Indeed, the lateral forces on a non-spinning baseball are very complicated and the tiniest difference in orientation or stiches can lead to very different forces. For this reason, both batter and catcher are unaware of the ball's final position in this situation. The last minute veering can happen when it is too late for the batter to coordinate a swing.

WHY DO BASKETBALLS BOUNCE BETTER THAN TENNIS BALLS?

Basketball bouncing.

When a ball bounces, it loses height, since only part of the energy is recovered after the rebound, the rest being lost in the impact.

We call coefficient of restitution (COR) the ratio

$$e = \frac{v_{after}}{v_{before}} \quad (e \text{ like « elasticity »}),$$

where v_{after} is the velocity after the rebound and v_{before} the velocity before.

The closer this parameter is to 1, the more elastic the rebound, and the less speed and height are lost. On the contrary, the closer this parameter is to 0, the greater the loss.

The proportion of velocity loss, *e*, is also reflected in the loss of initial height. Indeed, the height *h* is related to impact velocity v by the formula $v = \sqrt{2gh}$ (see previous chapter). So, if h_{before} is the height of the ball before rebound and h_{after} is the height after rebound, we also have

$$e = \sqrt{\frac{h_{after}}{h_{before}}}.$$

The coefficient *e* depends on the properties of the ground. A tennis ball or basketball does not bounce in the same way on concrete, clay or grass.

An approved tennis ball should bounce on a concrete floor from a height of 100 inches (254 cm) at a height of between 53 and 58 inches, i.e., 134.62 to 147.32 cm. Its coefficient of restitution should therefore be between 0.73 and 0.76.

In basketball, the homologation rule requires that, if the ball is dropped from a height of 1 m 80, it bounces back to a height of between 1 m 20 and 1 m 40, i.e., the coefficient of restitution must be between 0.82 and 0.88.

It is easy to check that a basketball does indeed bounce better and longer than a tennis ball. But even if we knew how to make tennis balls that bounce better, they wouldn't be approved.

In basketball, it is important to play fast, so the ball has to bounce quickly off the ground. But in tennis, if the bounce is too fast, serves would be even harder to return, as the speed of the ball would be very high. This explains the decision not to approve balls with a higher coefficient of restitution.

WHY SHOULD YOU OFTEN CHANGE YOUR GOLF BALL WHEN IT IS COLD?

The coefficient of restitution is also used to measure the effectiveness of a collision between a ball and the tool with which it is struck: golf club, baseball bat, billiard cue.

For example, the higher the coefficient of restitution with a golf club, the further the ball will go. The velocity acquired by the ball v_{ball} can be determined as a function of the club speed at impact v_{club}, the ratio of the masses of the ball and club, and the coefficient of restitution

$$v_{ball} = \frac{1+e}{1+\frac{m_{ball}}{m_{club}}} v_{club}.$$

This tells us that the faster the player's movement at impact, the greater the speed of the ball. What is important in golf, then, is not the force with which you hit the ball, but the club's rotational speed at impact (and the accuracy of your shot). The mass of a golf ball is approximately 46 g, the club 200 g, giving a denominator of 1.23.

Another important factor is the coefficient of restitution. In the 1970s, the coefficient of restitution of a golf ball was around 0.7, which meant that on impact at 115 mph, the ball acquired a speed of 1.7 × 185/1.23 = 159 mph. Since then, the technology used to manufacture balls and clubs has improved greatly, and the coefficient of restitution is now 0.86.

The movement of the ball also depends on temperature. It is reduced when the ball is cold. For example, it can drop by around 25% at 0°, and can fall down to 0.67. So, on winter days, it is a good idea to keep a ball in your pocket to warm it up, and to change balls regularly. Temperature affects other bounces, such as the bounce of shoes on a runway; the bounce of shoes also increases with temperature.

WHAT ARE THE SWEET SPOTS ON A TENNIS RACKET?

Tennis is not just about hitting the ball. It is a game of power and precision: you need to give the ball speed, but you also need to control its direction, without hurting yourself or triggering a pain in the elbow — the famous "tennis elbow". To achieve this, you need to know where and how to hit, and you do this differently depending on whether you are serving or returning the opponent's ball.

When serving, you want to transfer all the impact energy to the ball. Therefore, you should choose the point where the coefficient of restitution is lowest, i.e., all the energy goes into the ball. This is called the dead spot. Conversely, if you hit a ball that you receive, all the energy remains in the racket and is therefore the worst place to return a ball.

To return a ball, you need to hit it in the middle of the racket, usually called the sweet spot, but the precise location depends on the aim. When the ball hits the racket, it causes vibrations on the strings and then on the frame. The aim is to ensure that as little vibration as possible is transmitted to your arm, to avoid injury, and that as much energy as possible is released, so that the ball can be returned at high speed.

A racket's vibration can be precisely studied. There are several modes of vibration, and it is mainly the first mode for which you want to limit the effect. To do this, you need to hit

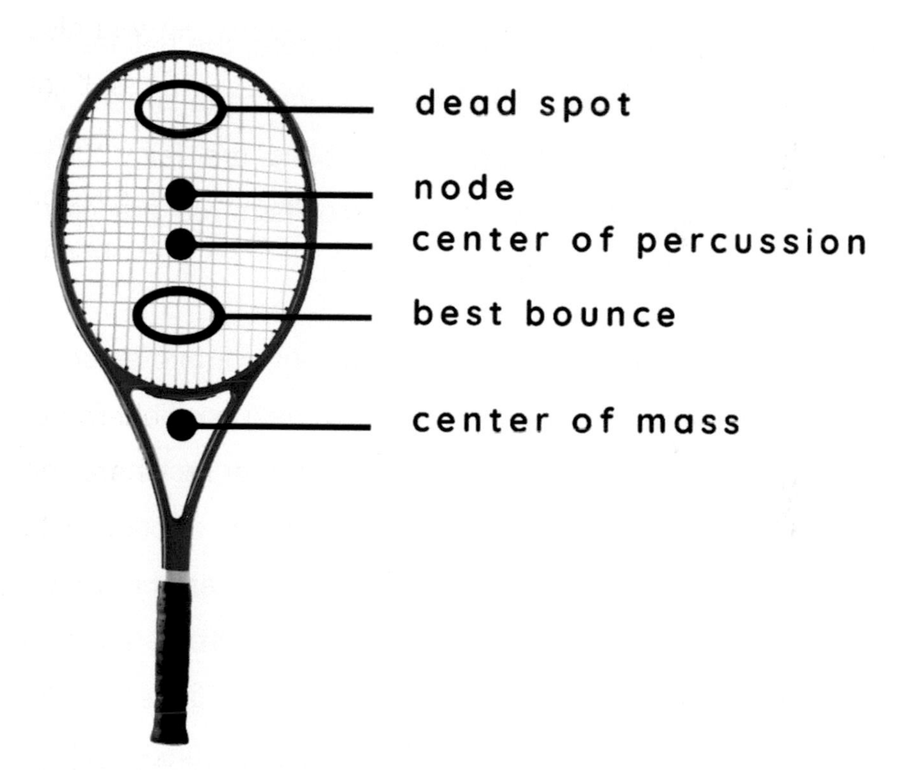

Sweet spots on a tennis racket. The dead spot for serving, the best bounce point for returning the ball. The racket's centre of gravity (due to the weight of the handle) is not on the sieve but a little lower.

the ball close to the node, a point that doesn't vibrate at all. Another important point is the centre of percussion, the only point on the racket that doesn't move at the moment of impact. When a ball hits a racket, the latter moves backwards (translation) and rotates around its centre of gravity (rotation).

The combination of these two movements leaves a single point fixed, called the centre of percussion. The last important point is the centre of oscillation, the area of the sieve where the coefficient of restitution is at its maximum and therefore where the rebound is maximized. From here, the ball leaves with the greatest speed. The best rebound zone is near the bottom of the racket. These three points are fairly close together, and it is difficult to hit any one of them with precision, so this is generally the area you aim at.

A great deal of research is being carried out to improve racket performance and help players return the best possible shots. Materials have evolved since the first wooden rackets. After being made of metal, today's rackets are made of composite materials, especially carbon fiber, which makes them very light. Strings, too, have changed greatly, both in terms of material and length. Longer strings (e.g., on large heads) have more deformation and will therefore transfer more energy to the ball, enabling more powerful play and faster balls.

In the 1980s, thicker frames were introduced, giving greater rigidity. The disadvantage is that it is less precise and often less comfortable, as vibrations are reflected back into the arm. On the other hand, these frames allow a much more powerful game. A rigid frame deforms less, so absorbs less of the ball's energy, which is then returned to the ball, which starts up again more quickly. On the other hand, a flexible frame deforms, absorbing the energy of the ball, which remains in contact with the string for a long time, and the effect imparted to it can be significant. So, in theory, softer rackets can slightly enhance the use of spin, for example, in lifts.

In the 1990s, long rackets were introduced, giving more speed as the point of impact is further from the pivot point, thus providing angular velocity, albeit to the detriment of precision for inexperienced players.

Technological progress has therefore made it possible to increase the playing speed of the greatest champions, a compromise having been found between power, precision and spin.

IN RUGBY, WHY IS IT BETTER TO STEP BACK TO CONVERT A TRY?

Goal kicking in rugby is a full body movement full of angles, balance, strength and timing. As all rugby specialists know, the best way to convert a try is to move backwards to open up the angle, even if this means taking a longer kick. But where and how should you move backwards? The facts are as follows: a rugby pitch is 70 m wide and the goalposts are 5.60 m apart. On a given vertical line, from point *A* with coordinates (x, y) (that is varying y), we need to find the maximum angle α at which we can see the goalposts. For simplicity's sake, we can assume that the poles are points *B* and *C* with coordinates *(-1,0)* and *(0,1),* which is equivalent to dividing all distances by the half of 5.6, i.e., 2.8. We want to express α as a function of x and y in order to optimize it.

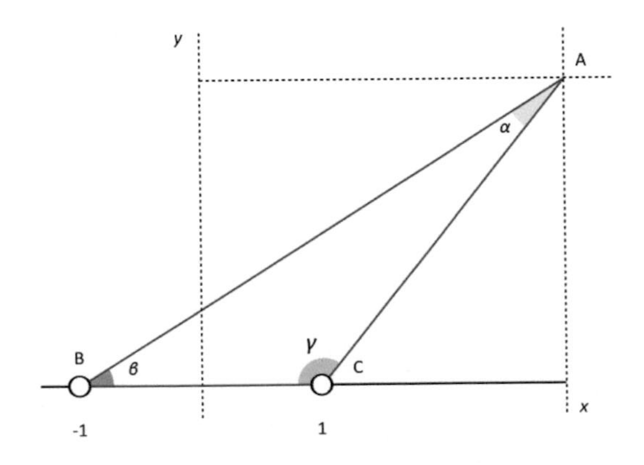

Point *A* of coordinates x and y where you kick. *B* and *C* are the goalposts.

The simplest way is to calculate the tangent of the angles. Recall that

$$\tan\alpha = -\tan(\beta+\gamma) = -\frac{\tan\beta + \tan\gamma}{1 - \tan\beta \times \tan\gamma}.$$

Moreover, $\tan\beta = \frac{y}{x+1}$ and, $\tan\gamma = -\frac{y}{x-1}$. This allows us to calculate

$$\tan\alpha = \frac{2y}{x^2 + y^2 - 1}.$$

A simple calculation of derivative yields that the maximal value of the tangent of α is reached on the curve $x^2 = 1 + y^2$. So, if the kicker is not facing the goalposts, it is better to step back at a distance $y = \sqrt{x^2 - 1}$. Given the distances in rugby, we are interested in the case where x is greater than 10, so we roughly have $y = x$ for the optimal kick. At this distance, the angle at which the player sees the goalposts is maximal. He must therefore be at the same horizontal and vertical distance from the centre of the posts. So, if he is at a certain distance *d* from the touchline, he needs to be at distance 35-d from the line of the goalposts and 35-d-22 from the line of the 22s! This time, it is not a question of the effect of air, but of maximizing the opening angle of the shot.

CHAPTER 4

WATER, WIND, COLD

WHY IS IT BETTER TO INHALE THAN EXHALE WHEN FLOATING?

To float in water, a body must be less dense than water, i.e., weigh less than if it were entirely filled with water. It is this difference in density that allows the body to float, thanks to Archimedes' buoyancy force: any object immersed in a liquid experiences an upward force equal to the weight of the displaced liquid. So, when an object is placed in water and floats, it displaces a volume of water equal to the volume of the submerged part of the object. At equilibrium, the total weight of the object is equal to the upward buoyant force, itself equal to the weight of the volume of water displaced by the submerged part. Given the density ratios of water and the human body (1000 kg/m^3 for the former, 950 kg/m^3, or 95%, for the latter), the human body floats. This is well known when wakeboarding. So, on average, 95% of a human body is immersed in water. Of course, this depends on the body — as muscle is heavier than fat, a very muscular athlete with no fat will float less than the average Joe with the same mass.

© The Author(s), under exclusive license to Springer Nature Switzerland AG 2024

A. Aftalion, *Be a Champion*, Copernicus Books,
https://doi.org/10.1007/978-3-031-54082-0_4

But there is a way to modify the force that makes you float. If you increase the volume of the submerged part without changing its mass, you increase the buoyant force and you float higher. A simple way to do this is to fill your lungs with air. The volume of your lungs, and therefore that of your body, increases, and since air is much less dense than water, the force pushing your body upwards, increases. On the other hand, if you breathe out very hard, you reduce the volume of your lungs, and therefore that of your body, and sink. This buoyant force also explains why, to go underwater, you have to exert a downward force, either with your arms or by weighting yourself down.

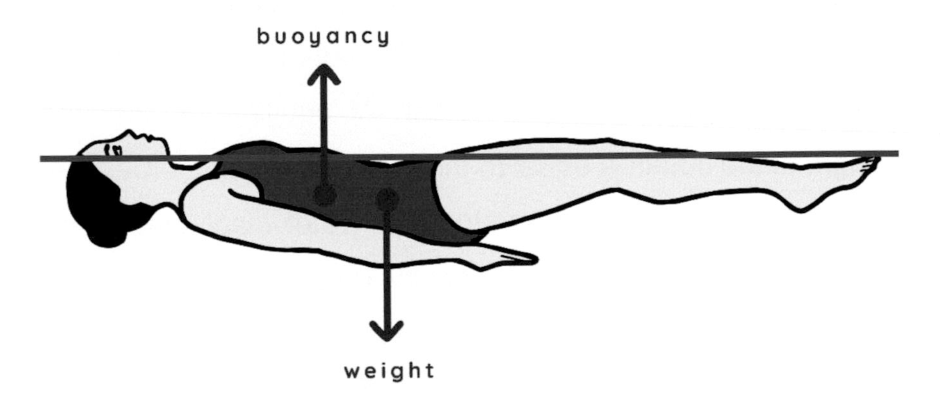

Centre of gravity and weight, centre of buoyancy and buoyancy force. The buoyancy force acts at a point to the left of the centre of gravity and will therefore cause the swimmer to turn clockwise.

Why can't we feel the force of lift in the air as we do in water? Simply because air, with a density of 1.2 kg/m^3, is much less dense, about 1,000 times less than water. The force of lift therefore also exists in the air, but the ratio of air lift to weight is generally very low, on the order of one to a thousand, and therefore negligible.

The buoyant force acts at the centre of buoyancy, which is the centre of gravity of the submerged part of the object. For the object to maintain its orientation in the water (i.e., not tip over), this buoyant force must also pass through the object's centre of gravity. If it is offset from the latter (see the figure above), the object will rotate until the centre of buoyancy and centre of gravity are aligned and rotational equilibrium is then achieved. Breathing in or out may cause the body to rotate to re-establish the balance of the centre of gravity, as it modifies the submerged part.

WHY DO YOU SWIM FASTER WHILE SLIGHTLY UNDERWATER?

Swimming involves an interaction between water and the swimmer. By moving his arms and legs through the water, the swimmer creates a force that propels him forward, but is slowed down by another force, the drag force, opposing the swimmer's movement through the water. The faster the swimmer moves, the greater this drag force. The drag must be reduced as much as possible to optimize the swimmer's effort. The drag experienced by a swimmer at the surface can be separated into three components:

1. Pressure drag — this is due to the fact that the swimmer pushes the water out of his way as he swims, creating a pressure difference between the water in front of him (high pressure) and the water behind him (low pressure). This drag is reduced if the swimmer is as horizontal as possible and splits the water; this reduces his frontal area in the direction of movement, thus minimizing pressure drag.

2. Skin friction — the water flows around the swimmer to let him pass, creating friction between the water and the swimmer's body as the former flows along the latter. This type of friction occurs within the very thin layer of water directly touching the body. It is to reduce this friction that swimmers shave and wear swim caps. Friction is greatly reduced by a wetsuit that reproduces

shark skin and glides through the water. These polyurethane wetsuits helped break a number of swimming records between 2008 and 2010, before being banned from competition.

3. Wave drag — this is caused by waves formed by arm and leg movements during swimming. At low speed, it is the friction against the body that dominates, but at the speed of competitive swimmers (around 2 m/s), it is the waves that create the biggest drag, so it is important to have a swimming technique that limits their amplitude.

These drag forces are reduced when the body is completely surrounded by water, i.e., submerged: on the one hand, the resistance created by the friction of water on the body is lower underwater than on the surface. On the other hand, the waves formed by arm and leg movements are reduced. Submerged, a swimmer's body moves faster, and encounters about two and a half times less resistance underwater than on the surface. So, when a child learns to swim, and has a natural tendency to do so slightly underwater, with roughly the equivalent of a body thickness above him, he spontaneously finds the optimal position.

The dolphin kick underwater swimming technique improves propulsion. The swimmer pushes the water (creating a thrust) with an undulatory motion, and according to Newton's third law, the water in turn pushes the swimmer, creating a thrust that propels him or her forward. This is how competitive swimmers start a race: they stay underwater, undulating, pushing alternately up and down. Swimmers can propel themselves at speeds of up to 2.5 m/s.

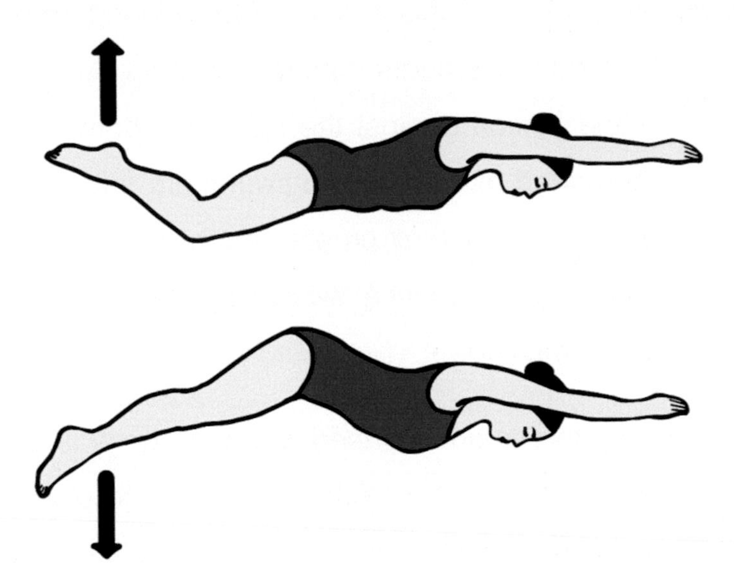

The dolphin kick. The swimmer whips the water upwards and then downwards while undulating, causing the water to push the swimmer away.

The effectiveness of the dolphin kick has forced the international swimming federation to restrict starts and turns to 15 m in competition at the risk of disqualification. In 2010, the American swimmer Hill Taylor accomplished a 50 m backstroke using the dolphin stroke, beating the world record of the time by one second. In today's competition, swimmers use these 15-metre authorized underwater undulation to optimize their performance.

WHY IS IT HARDER TO ROW OR PADDLE IN SHALLOW WATER?

Water depth has an effect on boat speed, regardless of wind or waves. Like a swimmer, a canoe or kayak is slowed down by a force of drag, made up of three components: shape drag, linked to the shape of the hull, viscous resistance on the hull, and wave drag. Boat hull shapes and materials are designed to limit this frictional force. The water forms a film around the hull, which moves forward at the same time. It is only in very shallow water, of the order of a few thicknesses of this film, that the resistance to movement of this part of the drag becomes very strong. It is not because you risk hitting the ground that you have trouble moving forward, but because the water no longer circulates around the hull to form this film that moves with it.

Wave drag, on the other hand, depends very much on depth, and becomes more significant in shallow water, when waves can interfere with each other to create resistance three to four times greater than in deep water. Take a 4 m 50 kayak travelling at 4.5 m/s: resistance is greatest at a depth of 2 m 55. When the ratio $v/\sqrt{gd}$, where v is the speed, g is the gravity, and d is the depth, is around 0.9, resistance is maximum. But this dependence on depth is highly complex. Resistance is lowest at a depth of 1.05 m, but then very high again at 80 cm, becoming very low again at around 8.30 m.

WHY CAN A SAILBOAT SAIL FASTER THAN THE WIND?

In ancient times, the sails of the first boats were square rig. You couldn't move forward without a tailwind, and you couldn't move faster than the wind. This is why the Greeks had to wait for favorable winds before going to war against Troy. In the Mediterranean, the wind patterns were too irregular to make do with a tailwind and a square rig, and the introduction of the triangular latin-rig or lateen in the 9th century made it possible to "sail upwind", i.e., to sail against the wind by changing the angle of inclination of the sail.

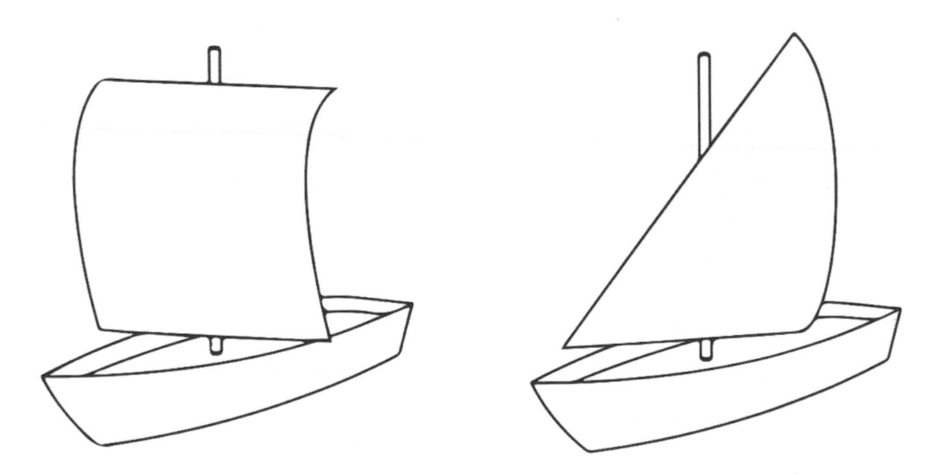

(Left) a square rig and (Right) a latin rig.

Today, you can see boats that are much faster than the wind, sometimes just with a crosswind. How is it possible?

The way a sail works depends on the direction of the ship relative to that of the wind. The figure shows the general case where the wind blows at a velocity V_{wind} with an angle θ to the horizontal. The flow along the sail creates a pressure difference between the windward side and the leeward side. The sail is oriented to optimize the flow of air around it and generate as much lift force as possible. The flexibility of the sails enables them to mimic the behavior of an aircraft wing, and to be oriented in many different positions to obtain the maximum propulsion, depending on the wind direction.

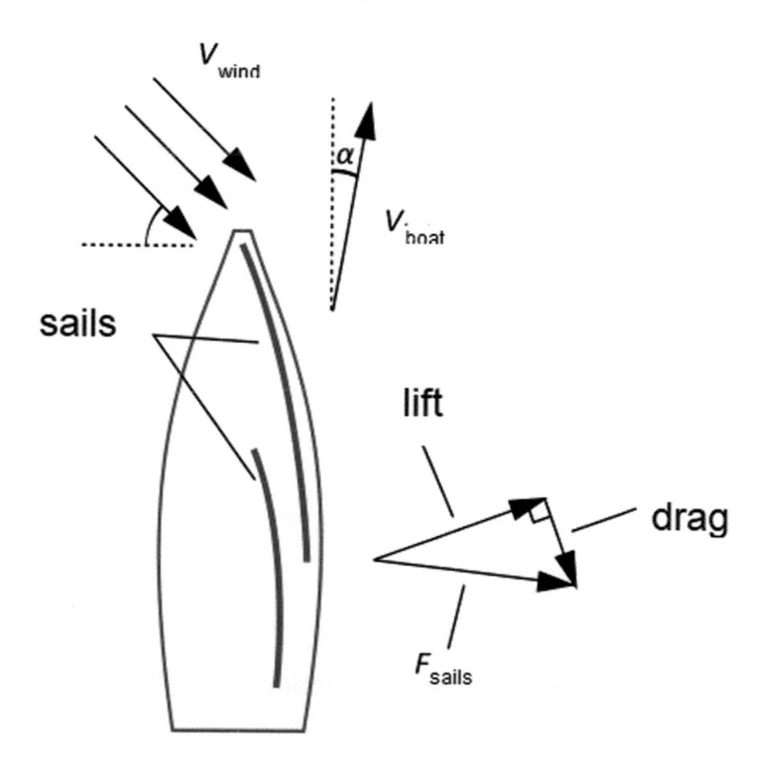

Lift and drag on a sailboat with wind.

The resulting force on the sails, called F_{sails}, is roughly perpendicular to the sail (more precisely, to the chord of the sail plan). It can be broken down into two parts:

1. lift, which acts perpendicularly to the wind direction, and is propulsive
2. drag, which acts parallel to the wind direction, tends to make the boat drift, but can also dangerously compromise its balance, or even cause it to capsize. This is where the keel or centreboard comes in.

The keel behaves like an underwater wing, and the same physics as the sail applies. The force exerted by the water is broken down into a lift force and a drag force. The role of the keel is to create as much counter-force as possible to the drag of the wind on the sail, in order to minimize lateral movements. When the forces are balanced, the sailboat moves at a constant speed V_{boat}, depending on the angle of attack that the keel makes with the currents. The direction of V_{boat} is slightly asymmetrical to the right of the boat's axis. The sailboat does not move "head-on" in the water; it is necessary for the sailboat to be slightly off-centre, as this allows the keel to generate the counter-force needed to resist the lateral force exerted on the sails by the wind. The two main components of a sailboat that enable it to move forward efficiently are the sail and the keel.

If the wind is blowing from the side, it is possible for V_{boat} to be greater than V_{wind}. This is because the lift force can be sufficiently strong. The sail exerts a propulsive force as long as its angle to the apparent wind (the difference between wind speed and boat speed) remains sufficiently large (of the order of 30 degrees). It is therefore possible to go faster than the actual wind. This is also the case for sailboards, multihulls and sand yachts.

If the boat is fitted with a foil, a specially shaped fin under the hull, it may even rise above the water. This is called foiling. The foil creates an upward force that increases with speed, compensating the weight and replacing buoyancy at high speed, lifting the boat above the water surface. The force opposing propulsion is then reduced all the more, and the boat can appear to fly above the water.

It is impossible for a sailboat to move directly into the wind ($\theta = 90°$ as in the figure above), because the resultant force F_{sails} has no lift component, so there is an upper limit to the size of the angle θ. For high-performance sailboats, this limit is around 60°. If, on the other hand, the boat is downwind, speed tends to reduce the apparent wind (the difference, remember, between wind speed and boat speed). So, contrary to intuition, this downwind situation is not the fastest, as it is not possible to go faster than the true wind.

WHY DOES A SPOILER PREVENT RACING CARS FROM TAKING OFF?

It may seem paradoxical, but movement on the ground is achieved thanks to friction. If there was no friction on a road, cars would slide and skid. It is friction that enables the wheel to turn instead of sliding, and thus to adhere to the ground. The force of friction opposes movement and is proportional to the reaction of the ground. This coefficient of proportionality is of the order of 0.01, i.e., the friction force is equal to 0.01 multiplied by the weight of the car on a flat road, so for a car of 1000 kg, it is 98 N.

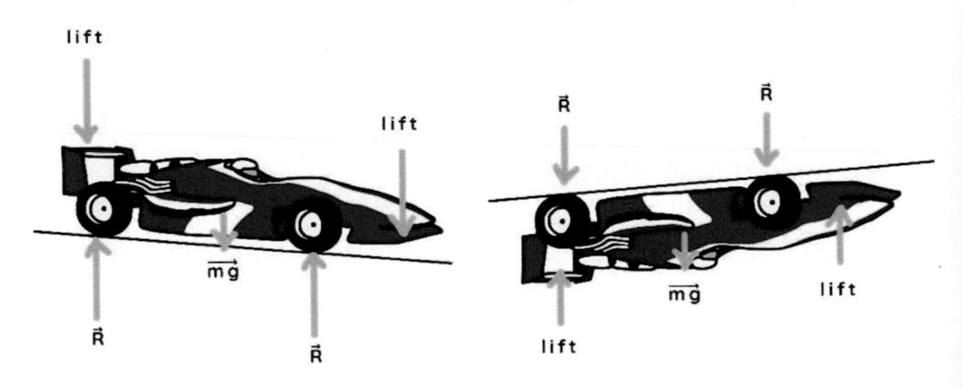

Forces on a card: weight, ground reaction and lift.

Friction is also necessary to prevent the car from taking off when on a bend, where centrifugal force would tend to tilt it. A simple way of increasing the frictional force is to increase the vehicle mass. The greater the weight, the greater the ground reaction, and therefore the greater friction. But this requires more fuel to move forward, so it is not a wise choice.

To increase grip without increasing mass, spoilers were introduced on racing cars. Spoilers create an aerodynamic force that presses the car downwards — the same force that acts on an aircraft wing, but in the opposite direction.

In this way, the car is subject to its own weight and the force of lift, both of which are directed downwards. The ground reaction, which must be equal to the sum of the two for the car not to lift off, increases for a car with spoilers.

The friction force therefore also increases, and the car does not lift off. At high speed (of the order of 200 mph), this lift force reaches a value of the order of three times the weight. If the road was upside down, the car could even continue to stick to it!

WHY SHOULDN'T YOU LOWER YOUR HEAD TOO MUCH ON A BIKE?

On a bicycle, the main force opposing movement is air resistance, also known as drag, which is proportional to the square of the speed, i.e., speed multiplied by itself. As cyclists sometimes reach speeds close to 35 mph, it is very important to find ways of limiting it. This resistance has three causes, as for swimmers:

1. air hits the cyclist at the front,
2. air flows around him to let him pass,
3. eddies form in the cyclist's wake.

To reduce drag, a cyclist must

- adopt an aerodynamic position to reduce frontal impact,
- choose an equipment (helmet, clothing) that allows air to circulate around him. He will even shave the parts of his body which are not protected by clothing, to help the air slide better,
- and reduce the vortices in his wake as much as possible.

In addition to the helmet, its shape and technology, the position of the head is also important. A helmet with too much or too little head angle can increase drag and slow down the athlete! Let us take a closer look.

Air effect on two cyclists behind one another.

On a bike, it is better to be at the back than at the front: inside a peloton, you can maintain your speed without too much effort, feeling as if you're being sucked along, whereas the leader at the front feels tired after a few minutes. This is because behind any object moving through the air, there is a wake in which the pressure is lower. So, if cyclist number 2 follows his leader closely enough, he finds himself in a low-pressure zone. Not only is there no air coming at him, but he also benefits from a suction effect as shown in the figure.

It is less well known that the leader also benefits from a suction effect, reducing his drag by around 5%. Indeed, cyclist 2 occupies his wake, which reduces the vortices behind cyclist 1, and thus reduces the drag that opposes movement. But this only works if cyclist 1 adopts an aerodynamic position.

The same effect occurs with the helmet. The helmet is designed to allow air to circulate easily around it when the head is raised, but if the head is too low, a wake of vortices forms behind it and drag increases. The helmet is supposed to take up the space behind the head to reduce this wake. This type of helmet is not recommended for the Tour de France, where the rider has to look both at the bumps and dips in the road and at the competitors behind him.

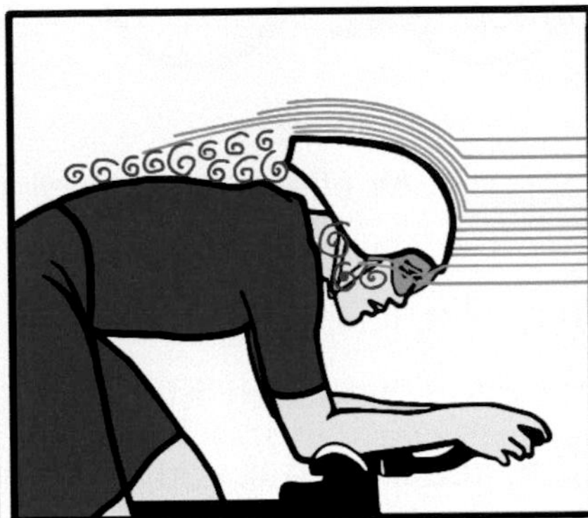

Air effect on the helmet according to the head position. On the left, the air flows properly, on the right, the head is too low and vortices slow down the movement.

All this is independent of the wind and is only linked to the fact that a cyclist is going fast and splitting the air as he goes. Without wind, the optimal position for cyclists is the pace line as the velocity at which air collides with them is the opposite of their velocity.

Now, let us imagine that there is wind. If it is exactly in the opposite direction to the movement, it is similar to the no wind situation as the wind speed is added to that of the cyclist, because in a frame of reference where the cyclist is fixed, the air speed is equal to the cyclist's actual speed plus that of the wind.

If, on the other hand, the wind is transverse, that is perpendicular to the movement, cyclists must not ride one behind the other, but offset diagonally. The angle of this offset is calculated as $\tan\theta = \frac{v_{wind}}{v_{bike}}$. This is reminiscent of the position of migratory birds in the sky, which choose an angle corresponding to the wind.

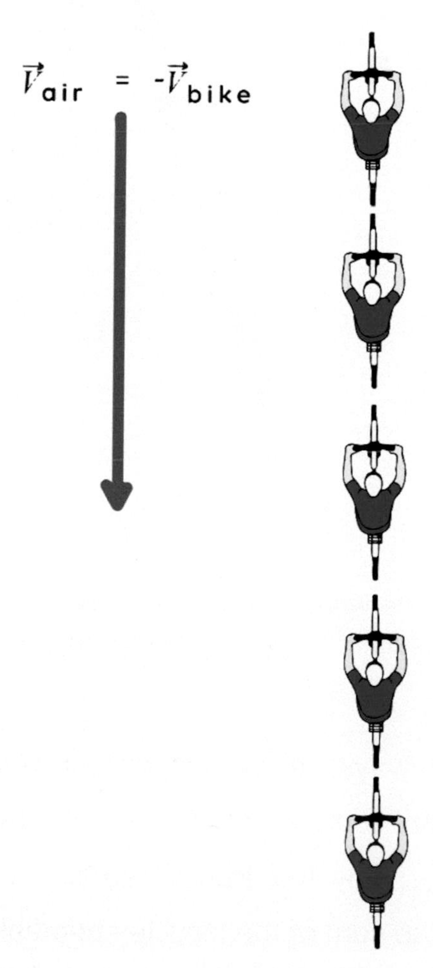

If there is no wind, the velocity at which air hits a rider is the opposite of the rider's velocity. The optimum position for a group of riders is a pace line.

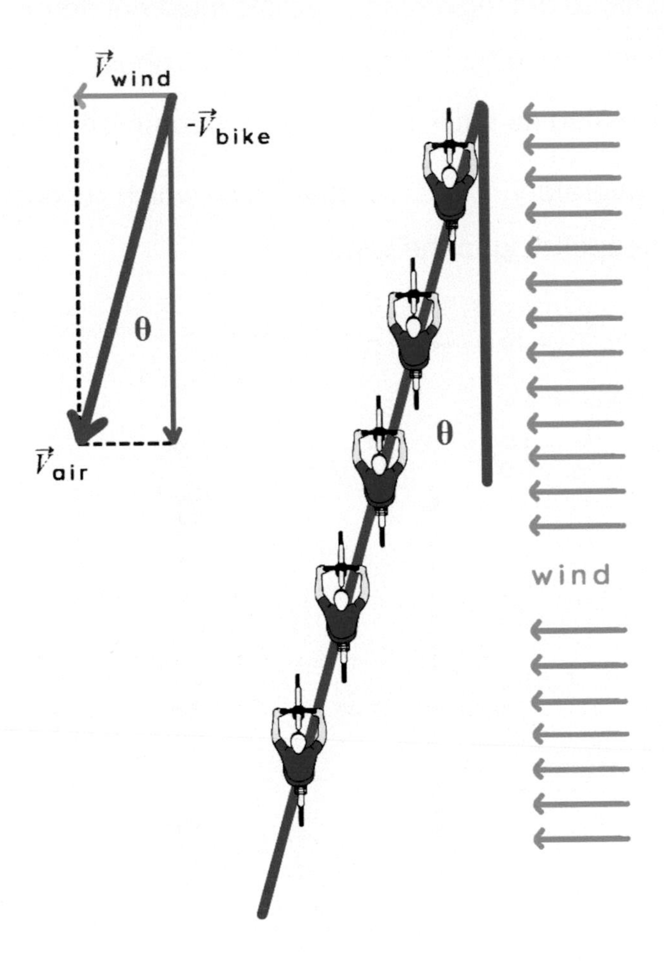

If there is a side wind, the velocity at which air hits a rider provides the pace line. It is the sum of the wind velocity and the opposite of the rider velocity.

If the wind is sideways, it therefore has a transverse component and a front component. We need to replace in the tangent formula, v_{wind} by the transverse part of the wind speed and v_{bike} by the sum of the bicycle speed and the front wind speed.

WHY ARE VELODROME TRACKS HEATED?

Cyclists on velodromes go very fast and are therefore subject to very high drag, the resistance of the air opposed to movement. To reduce drag, cyclists adopt an aerodynamic position. But the most effective way to reduce drag is to be in a place where the air density is lower (drag is proportional to the number of collisions and therefore to the air density). To achieve this, there are two solutions: gain altitude or increase temperature. Indeed, the higher you are, the smaller the density. At the 1968 Olympic Games in Mexico City, many throwing and jumping records were broken, and it is thought that the altitude of 2,240 m and the reduction in drag it induced may have been responsible. For endurance sports, on the other hand, the reduction in air density is unfavorable, as there is less oxygen.

What about temperature? There are 13% fewer molecules in the air at 37° C than at 0° C, which improves golf ball distance by 7%. The same goes for cyclists. Drag is reduced at higher temperatures. That is why the London velodrome track for the 2012 Olympics was heated to 28° C. This was not very pleasant for the audience, and it would have been better for them if it had been 20° C, but between 20° C and 28° C, a cyclist's drag is reduced by 3%, which is quite significant for performance.

WHY CAN A DOWNHILL SKIER GO FASTER THAN A SKY DIVER?

One might think that maximum speed is achieved by a free-fall jump from a sufficiently high altitude. The world record for freefall from space at 40,000 meters is indeed 830 mph, but those who jump from high mountains, at 3,000 or 4,000 meters, fall at speeds of the order of 125 mph. And skiers can go faster on the ski run. Indeed, their speed records are of the order of 158 mph. Why is this so?

Skydiver : side view and bottom view of the spread eagle position.

Before opening his parachute, a skydiver is subject to his own weight, which causes him to fall, and to air resistance, which slows him down and is proportional to the square of his speed v $(v \times v)$ and the surface area A he presents to the wind, multiplied by the proportionality coefficient κ. In general, skydivers assume the spread eagle position, with arms and legs spread out and slightly bent (see the figure above).

During freefall, the skydiver reaches a speed where the two forces, weight and air resistance, balance out so that

$$mg = \kappa A v_{skydiver}^2.$$

The speed skier crouches in a tuck position in order to present to the air the smallest possible area, which is much smaller than the skydiver's (see the figure below). For the skier, speed analysis is more complicated, as it depends on the angle of the slope θ. Air resistance is of the same form $\kappa A' v^2$ as the skydiver, but because of the different position, A' is of the order of half A.

In the aerodynamic position, the skier gathers his limbs as close to his body as possible: knees bent, chest inclined, back rounded, poles shaped to fit his body, he thus adopts a tuck position that offers little resistance to the wind.

To understand how the skier can go faster than the skydiver, we need to imagine that he has reached and maintained a constant speed, where forces balance out. We are interested

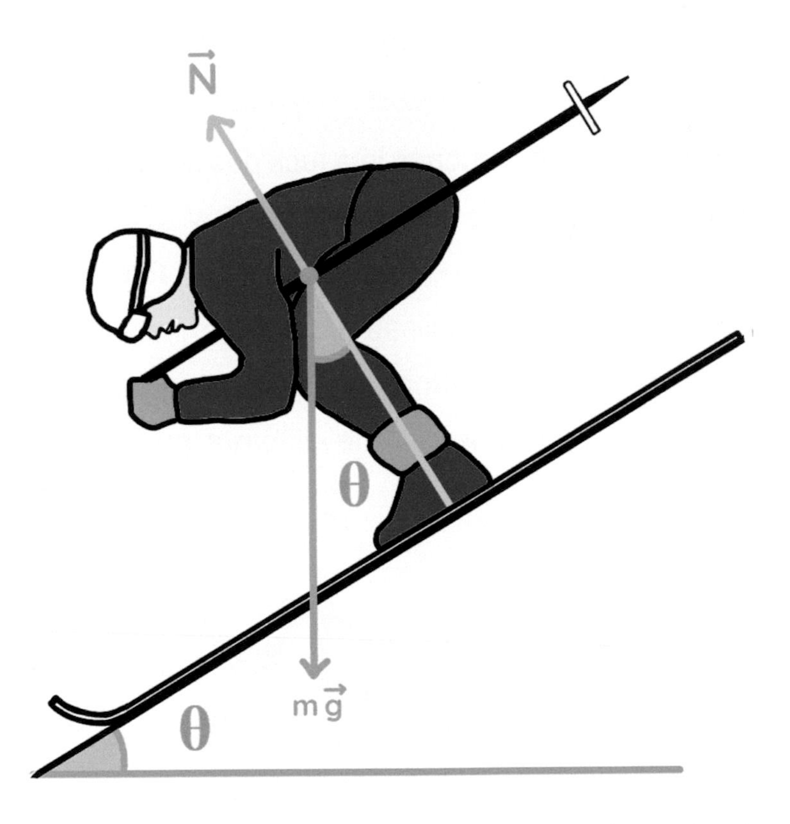

Downhill skier. The weight is vertical and the ground reaction is perpendicular to the ski run, with the same angle to the weight as the angle of the slope.

in the forces acting in the direction of motion (figure above). There is the weight, which pulls downwards, but with modulation by the angle of the slope, or more precisely by its sine, and there is the friction of the snow, which is more or less important depending on whether it is dry or wet. We call this coefficient of friction μ and it can vary from 0.04 for dry snow at 0 degrees, to 0.20 for wet snow. The resistance of the snow is proportional to the reaction of the ground via this μ coefficient, which is linked to the vertical component of weight when the skier doesn't take off.

This leads us to a value for the speed that is given by $mg(\sin\theta - \mu\, cos\theta) = \kappa A' v_{skier}^2$. We can therefore see that the ratio between the skier's speed and the skydiver's is

$$\frac{v_{skier}}{v_{skydiver}} = \sqrt{\frac{(\sin\theta - \mu\, cos\theta)\; A}{A'}}.$$

We have seen that $A/A' = 2$. The skier is therefore going faster than the skydiver (the ratio is greater than 1), if for example $\mu=0.09$ and the angle exceeds 35 degrees. The skier's speed is then 53.65 m/s or 120 mph. This is well below the record speed of 158 mph! This is even more true on very dry snow, which allows smaller slope angles.

WHY SHOULD DOWNHILL SKIERS AVOID JUMPING TOO FAR?

When the downhill slope breaks and suddenly becomes steeper, the skier cannot keep up. She has to jump! If she does not anticipate the jump and lets her weight carry her off, she loses precious time, because the jump is long. The longer the jump, the less able she is to maintain an aerodynamic position in the air, as the force of air resistance is greater and slows her down more than in a descent position. Above all, the longer the jump, the greater the force exerted on the legs on impact with the ground, slowing her down. Let us take a closer look.

At Val d'Isère in the French Alps, the slope for the women downhill course is around 15 degrees and the speed of the skiers is around 27 m/s (2300 m descent in 85 s). Let us imagine a skier who doesn't anticipate her jump when there is a slope discontinuity in the ski run. She is then subject to her own weight, and we can calculate the length of the jump (14m), and the evolution of her speed. At the end of the jump, she reaches a vertical speed of 12 m/s. The fact that there is a break in slope that increases to 20 degrees gives a link between the vertical velocity v_y and the horizontal velocity v_x at impact, which is $v_y = v_x \tan 20$. During the jump, the horizontal velocity does not change. On the other hand, the vertical velocity undergoes a variation of 2.5 m/s. This variation makes it possible to calculate the force the skier receives in her legs, equal to the variation in speed multiplied by the mass and divided by the time of the order of 3,000 N — or about six times her weight (note that the time is estimated as a function of the vertical speed and the difference in height of the centre of gravity, which can be estimated at 50 cm).

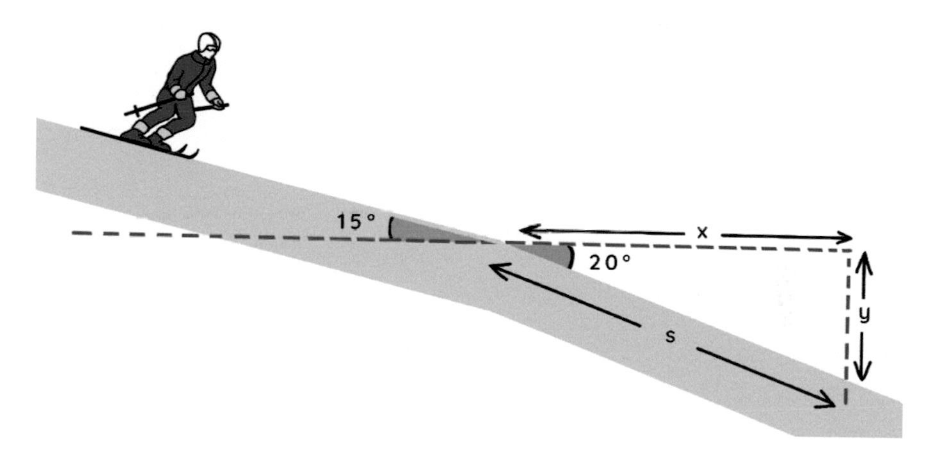

The slope of the ski run changes abruptly from 15 to 20 degrees. *s* is the distance along the slope, *x* is the horizontal distance from the slope discontinuity and *y* the vertical distance.

In order to stay in the air for as short a time as possible at the sight of a bump, she quickly gets up from her crouch position and pulls her legs upward. Therefore, she jumps in advance of the edge of the steeper slope. The time saved is real when the skier executes a short, cushioned jump. The timing of this cushioned jump is crucial, and can save the most skillful skiers almost two or three tenths of a second. If it is not right, the jump will be too long and too high. What sets an excellent skier apart is her aerodynamic position. The best skiers can sometimes recover as much as 7 hundredths of a second during take-off, and a flat, simultaneous landing on both skis is crucial to re-establishing contact with the snow as soon as possible. Moreover, the landing force on the skis is reduced. Anticipating an aerodynamic posture from the landing is an integral part of a jump.

CHAPTER 5
ROTATING

WHY ON A BIKE, THE FASTER YOU GO, THE MORE STABLE YOU ARE?

We have all experienced the difficult beginnings on a bicycle once you remove the little training wheels: a twist of the handlebars to the right, a twist of the handlebars to the left, balance is unstable and you are on the verge of falling over. And yet, if you are able to speed up, balance seems easier, and handlebar movements to the left and right diminish. Conversely, if you try to go very slowly, or even keep your balance at a standstill, it seems very complicated. Why does this happen? Why is it that equilibrium is established at the very moment when you start riding, and why can a jerk of the handlebars restore it?

The only two external forces acting on the bike + rider system are the rider's weight and the reaction of the ground on the bike's wheels.

- Weight is characterized by its vertical downward direction. It passes through the system's centre of gravity, which depends on the rider's position (see centre of gravity in the **Forces** box). It is therefore

© The Author(s), under exclusive license to Springer Nature Switzerland AG 2024

A. Aftalion, *Be a Champion*, Copernicus Books,
https://doi.org/10.1007/978-3-031-54082-0_5

important to understand where the centre of gravity is located.

- When a bike is on flat ground, weight and ground reaction exist on the same vertical plane. Ground reaction, characterized by its vertical direction, is directed upwards, and applies to the contact points between the wheels and the ground. As your bike has two wheels, the ground reaction is divided into two contributions, the sum of which is equal to the value of the weight.

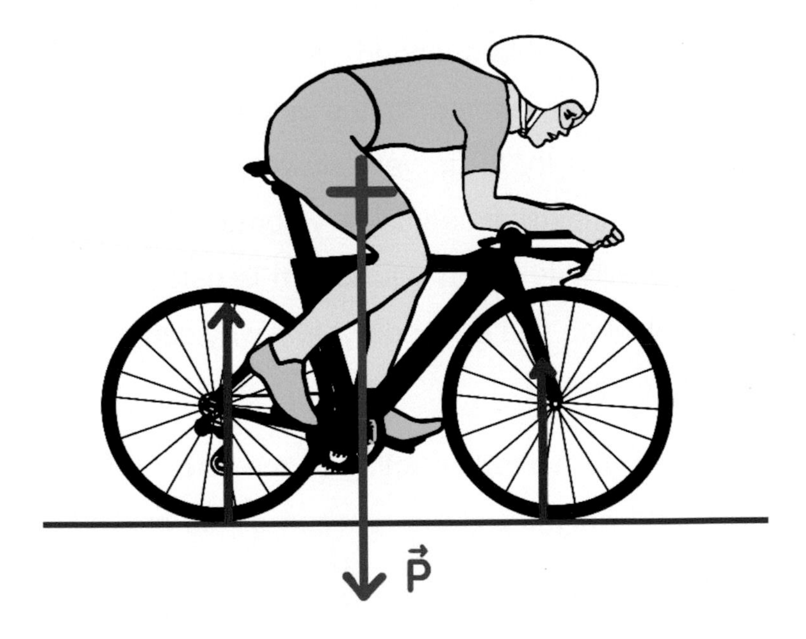

Forces on a bike and a cyclist: weight and ground reaction on the wheels.

For the bike to be balanced, the centre of gravity must remain above the area in contact with the ground (support base). As the area in contact with the ground is very narrow, the slightest shift in the centre of gravity causes a lateral disequilibrium in the system which is difficult to control. On the other hand, the wider the tires, the wider the support base, making it easier to align with the centre of gravity.

Cyclist standing on his bike, with the handlebars turned, is more stable.

With the handlebars turned, you increase the surface area on the ground, which can guarantee balance even when stationary. And if you stand up, you raise your centre of gravity, making it even easier to balance!

Torque

A torque is a measure of how much a force acting on an object causes it to rotate. Its magnitude depends on the force but also on the lever arm, i.e., the distance between the axis of rotation and the force. If the force passes through the axis, it has no effect on rotation, and the lever arm and torque are zero. If this is not the case, the torque is the product of the force and the lever arm.

The sign of the torque depends on the direction in which the force causes rotation: positive clockwise, negative counter-clockwise. The torque therefore depends not only on the force, but also on its point of application. If an object does not rotate, or rotates at constant speed without deforming, Newton's first law of angular momentum indicates that the sum of the torques is zero.

The notion of imbalance is associated with the rotational force exerted by the weight in relation to the axis of inclination of the bicycle/cyclist system. Following a slight imbalance, for example to the right, the weight creates a torque to the right that tends to cause the system to fall: $M = mg\, d \cos\theta$ where m is the mass of the system, d the height of the centre of gravity at equilibrium, g the gravitational constant, and θ the angle with respect to the vertical. The systems' equilibrium can therefore only be ensured by a torque in the opposite direction and equivalent to M. This is why turning the wheel generates, according to the action-reaction principle, a rotation of the system that compensates that of the weight. This rotation creates a centrifugal force giving rise to a torque in the opposite direction to that of the weight: $-m\, \frac{v^2}{R} d \sin\theta$, where v is the system speed and R is the radius of the trajectory. Note that R is related to the angle of inclination of the handlebars, and the smaller the angle, the greater R.

Bicycle rebalancing is therefore linked to handlebar rotation by the formula

$$\frac{v^2}{R}\tan\theta = g.$$

At low speed, the radius of the trajectory must be small, and if you want to turn in a very small circle, you have to give a big push with the handlebars, whereas at high speed, the radius is large, you are almost straight, and the trajectory modification is infinitesimal. The rider transmits force to the handlebars to turn the front wheel, thus controlling the inclination and preserving balance. At high speed, even a small

Cyclist inclined at an angle θ. Weight causes tipping.

steering angle causes a rapid lateral displacement of the points of contact with the ground; at slower speed, larger steering angles are required to achieve the same effect just as quickly. As a result, it is generally easier to maintain balance at higher speed.

The further forward the combined centre of gravity of the bike and the rider is, the less lateral correction the front wheel needs to maintain balance. Conversely, the further back the rider's centre of gravity is, the more lateral correction the front wheel has to apply, or the faster the bike has to go, to regain balance. On some recent bikes, where the rider is lying down, the centre of the bike/rider system is towards the front; stability is better. Cyclists are advised to carry their loads low on their bikes, in particular by letting their bags hang down on either side of their luggage racks. The torque for straightening will be less important if the weight is close to the fulcrum, in contact between the tires and the ground, as the lever arm is then smaller and requires fewer handlebar strokes.

Other technical elements, such as frame design, also play a part in the bike stability. The so-called trail is the angle formed by the straight line extending from the fork and the vertical line passing through the point of contact of the front wheel with the ground. It plays an important role in the bike handling and balance. The greater the trail, the more stable the bike. Racing bikes have a large trail, because their aim is to maintain a trajectory, and therefore stability, by turning the handlebars as little as possible.

Trail of a bike: distance between the steering axis and where the wheel touches the ground.

At low speed, a tilt of the frame causes the front wheel to rotate, thanks to the torque created by the reaction of the ground in relation to the fork axis. One consequence of the trail, which has a direct influence on maintaining a linear trajectory, is that the greater the trail, the greater the torque of frictional forces opposing any handlebar rotation. Conversely, the smaller the trail, the more maneuverable the bike, since less friction will oppose handlebar rotation, as is the case with mountain bikes. A low trail increases maneuverability, while a high trail makes it easier to make up for steering errors and avoid turning the handlebars at high speed.

As Einstein said, "Life is like a bicycle: you have to keep moving to keep your balance". (Photo by Reuters/courtesy of Archives-California Institute of Technology/Handout)

A bike balance therefore depends very much on the rotation of the handlebars. The faster you ride, the better the balance, because a slight tug on the handlebars will straighten you out. On the other hand, if you try to move a bicycle forward on its own, it will only do so for about ten seconds at most, and will eventually fall over. This is why riding a bike without your hands is so complicated, because you have to go extremely straight and in such a way that the handlebars do not turn on their own, risking tipping over.

WHY DO BICYCLE EFFORTS DEPEND ON THE SPROCKET?

To move a bicycle forward, you push on the pedals which turn a toothed chainring. Thanks to the chain, this chainring rotates the rear sprocket, also toothed, which is attached to the rear wheel and make it turn. Early bikes had only one size of chainring and sprocket. Today, there are more choices.

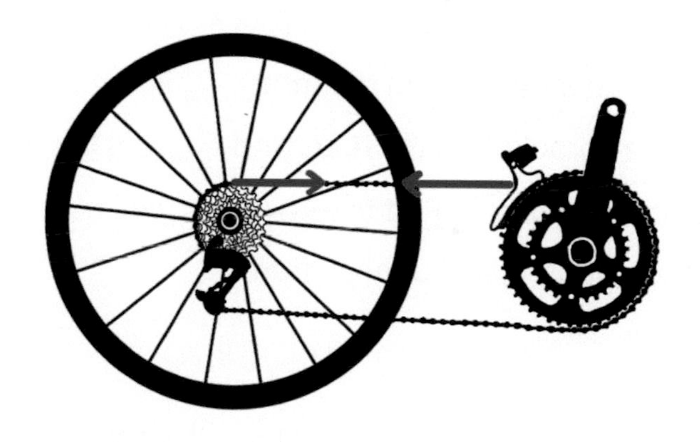

Sprocket at the rear and chainring.

Specialists know that for the same speed a large chainring requires greater pedal force with a lower cadence. On the other hand, a small chainring allows you to pedal more easily, but you have to increase your pedaling frequency to achieve the same speed. When the chainring makes one turn, it rotates its teeth (for example, 32) on the rear sprocket. If the sprocket has fewer teeth, it makes several turns. For example, if it has 16, one pedal turn will cause two wheel turns. What is the point of having more teeth on a sprocket? The smaller the sprocket, the greater the distance covered by one pedal revolution, and therefore the greater the pedaling force required.

The gear ratio is the ratio of the number of teeth on the chainring to that on the sprocket. It affects pedaling speed as a function of frequency. Let us try to understand this with a few formulas. First, all the points of the chain turn at the same speed, so if R_{re} is the radius of the sprocket (at the rear), R_{fr} is the radius of the chainring (at the front), f_{re} is the frequency of rotation of the sprocket (at the rear) and f_{pedal} is the frequency of rotation of the chainring (at the front), as well as the pedals, i.e., also the pedaling frequency, then

$$2\pi f_{re} R_{re} = 2\pi f_{pedal} R_{fr}.$$

By the principle of action and reaction, the force exerted by the chain on the chainring is the opposite of the force exerted by the chain on the sprocket.

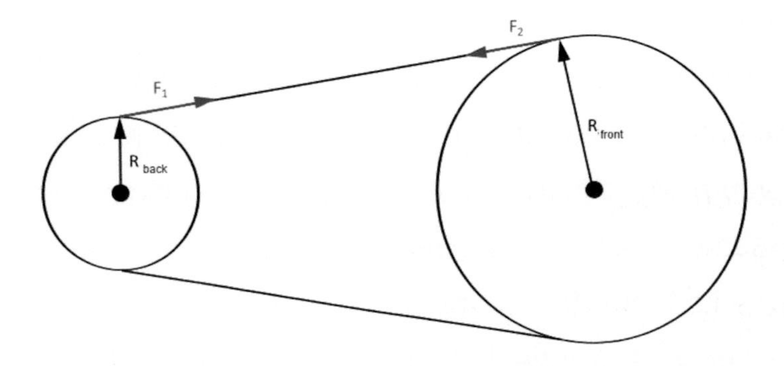

Schematic sprocket and chainring with forces.

Since gear ratio is the ratio of the number of teeth on the chainring to those on the sprocket, it is also the ratio of the radius of the front wheel to the rear wheel, and we know from the previous relationship that it is equal to

$$GR = \frac{R_{fr}}{R_{re}} = \frac{f_{re}}{f_{pedal}}.$$

Clearly, with a high gear ratio (e.g., 53/11), the frequency of the rear wheel relative to the front wheel is very high, since the rear wheel, integral with the sprocket, has the same frequency as the sprocket. Bicycle speed is thus given by the following formula, where R_{wheel} is the wheel radius,

$$v = 2\pi f_{re} R_{wheel} = 2\pi R_{wheel} \times GR \times f_{pedal}.$$

So the higher the gear ratio, the faster you go. But beware: it obviously takes a great deal of effort to maintain the same pedaling frequency when increasing the gear ratio! In general, cyclists are advised to find their optimum pedaling cadence and adapt the gear ratio according to the effort required and the target speed. Large gear ratios, e.g., 53/11 (= 4.8), with a cadence of 111.4 rpm, enable you to reach speeds of 42.2 mph, like Mark Cavendish in a flat sprint. On the other hand, on hills, cyclists choose a smaller gear ratio, e.g., 34/29 = 1.17, like Jérôme Pineau on the Monte Zoncolan climb in Italy, where he reached 3.7 mph.

But it is not enough to have a big gear to go fast, you also need a high pedaling frequency. By the principle of action and reaction, the force exerted by the chain on the chainring is the opposite of the force exerted by the chain on the sprocket. You need to understand how much force is exerted on the front chainring for the same effect at the rear. For a given gear ratio and pedaling cadence, the effort exerted by the cyclist is linked to what we call the torque of the force on the chainring, i.e., the product of the force and the distance from the axis of rotation. The greater the distance from the axis (i.e., the chainring), the greater the torque, and therefore the greater the effort. Larger chainrings are therefore preferred by more experienced cyclists. So why not use a small chainring all the time? The reason is because it generally creates a small gear ratio, and therefore a low speed. If you want to reach a high speed on the flat, you will choose a large chainring and a small sprocket which create a large gear ratio. It is all about finding the right balance between the speed you want and the effort you can put in.

On the other hand, for the same effort, i.e., the same torque, the large chainring requires a lower cadence because the lever arm is longer. But in return, you will go slower, since a lower cadence reduces speed. This is why experienced cyclists are able to go fast with large chainrings, which requires greater effort!

WHY CAN A GYMNAST TURN AROUND A HORIZONTAL BAR?

During a backward giant circle, gymnasts turn around a bar, arms outstretched, body more or less straight, and perform a complete rotation. This often takes place just before the dismount, in the form of a vault or somersault. The gymnast accelerates on the way down and slows down on the way up; she does not rotate at a constant speed. So there is angular acceleration and deceleration. It is her weight that makes the gymnast turn on the way down: the weight is applied to the

A gymnast rotating around a bar: backward giant circle.

centre of gravity, which is not on the axis of rotation, and therefore exerts a torque of value $mgl\sin\theta$, where m is the athlete's mass, g is gravity, l is the distance between the centre of gravity and the tips of the arms, and θ is the angle in relation to the vertical (see **Torque** box). The torque of the weight thus changes value, since it is zero when the athlete is vertical (head up or down) and maximum when she is horizontal. As we have seen, what counts is the lever arm. So, if the gymnast changes position, by bending her legs for example, she will modify the position of the centre of gravity (see the **Centre of gravity** box) and therefore the length of the lever arm.

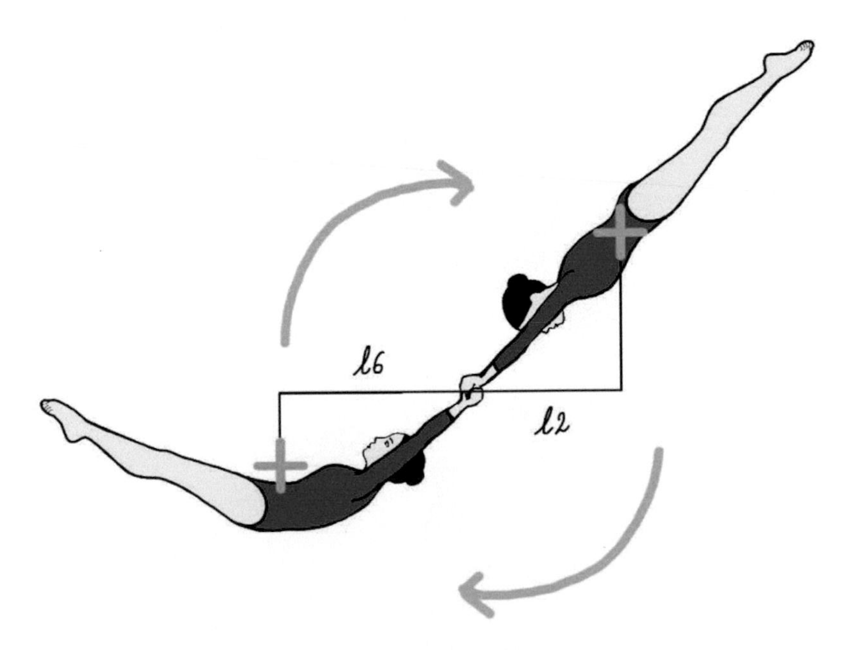

Two opposite positions. To come back up, the gymnast bends her legs to bring her centre of gravity closer to the axis and reduce the weight torque that is holding her back.

On the descent, where her weight accelerates her, she remains upright, but on the ascent, where her weight slows her down, she bends her legs to bring the centre of gravity closer to the axis and reduce the torque of the weight braking her. In technical terms, this is called a shoot.

WHY DO YOU PULL IN YOUR ARMS TO SPIN WHEN ICE SKATING?

To spin properly, you must first determine your balance position. The centre of gravity must be on the axis of rotation (see the centre of gravity in the **Forces** box). If you lift your leg backwards, you move the centre of gravity towards the back of your body. So in order to keep it more or less in its original position, depending on the axis of rotation, you need to lean your torso forward, as a skater does during an arabesque or a camel spin. If, on the other hand, you are extending your leg forward for a sit-spin, you need to shift your centre of gravity backwards, putting as much of your body weight as possible on the back of your foot. And to turn quickly and well, you mustn't wiggle, otherwise you will lose your balance. Or, if you do wiggle, you must re-establish the centre of gravity by shifting the shoulders to the opposite side.

The two forces acting on a skater spinning are weight and the reaction of the ice on the skate. Ground friction and air resistance, even in cold weather, can be negligible. As both forces apply to points on the axis of rotation, angular momentum is conserved. But speed can change when you change position. This is what we observe between a sitting spin (fast) and a camel spin (slow). It is all about the moment of inertia!

Spin: the centre of gravity has to be on the rotation axis.

Thanks to her momentum, the skater accumulates angular momentum, which is conserved during the spin: $I\omega$ where ω is the number of revolutions per minute and *I* is the moment of inertia (see **Moment of inertia** box). To maximize the speed of rotation ω, we need to minimize *I* and keep the body as close as possible to the axis of rotation. That is why skaters squeeze their arms as tightly as possible or raise them above their heads to accelerate, as shown in the next figure. They also accelerate when they bend their knees until they sit down in seated spins, because the body is more closely grouped near the axis of rotation.

Moment of inertia

The moment of inertia expresses the difficulty of rotating a body, taking into account the distribution of mass according to its distance from the axis of rotation. Indeed, a skater turns faster if she pulls in her arms and legs rather than spreading them. This is not due to air friction, but to the fact that the closer the mass is to the axis, the easier it is to rotate. If the body is cut into small pieces, the moment of inertia *I* is the sum of the masses of the pieces multiplied by the square of their distance from the axis. The lower the moment of inertia, the easier it is to rotate an object.

The further the mass is from the axis, the greater the moment of inertia (as in the case of the trapeze artist). But for a given position, the moment of inertia depends on whether the axis is horizontal or vertical.

Here are some values of the moment of inertia for an athlete:

Folded position in rotation around the centre of gravity. $I = 3.5\ kg \cdot m^2$	Semi-folded position rotated around the centre of gravity. $I = 6.5\ kg \cdot m^2$

Elongated body, rotating around the centre of gravity. $I = 15\ kg \cdot m^2$	Elongated body, rotating around a bar passing through the hands. $I = 83\ kg \cdot m^2$

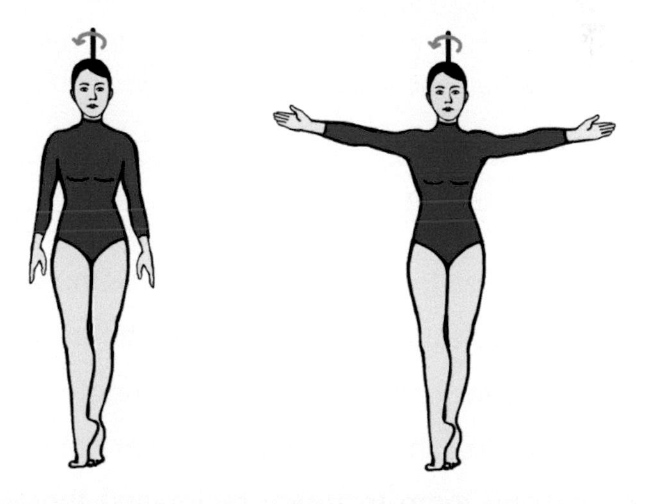

Standing position, arms at side, rotating around vertical axis. $I = 1.1\ kg \cdot m^2$	Standing position, arms spread, rotating around vertical axis. $I = 2\ kg \cdot m^2$

Angular momentum or kinetic momentum

For a rigid body rotating around an axis of rotation, angular momentum is the product of its moment of inertia I (which measures an object's resistance to being rotated) and its angular velocity ω. Angular velocity is the change in angle per unit of time, and is related to angular frequency (the number of revolutions per unit of time) by a factor 2π.

Newton's second law of rotation states that the variation in angular momentum of an object is equal to the sum of the torques of the forces to which it is subjected. If the sum of the torques is zero, for example if the forces all pass through the axis of rotation, the angular momentum is conserved. Otherwise, this law can be used to calculate the variation in angular momentum.

If an athlete performing a rotation for which angular momentum is conserved changes the position of his body, he also changes his moment of inertia I, and therefore his angular velocity. Simply by changing his body position, an athlete can change his rotational speed.

The same applies to ice skating jumps. Some young Russian skaters manage to do four rotations in the air with their arms stretched above their heads. The moment of inertia in this position is smaller than when the arms are at the sides of the body, enabling them to rotate faster and therefore perform more turns in the air in the same time as their competitors.

During all of their arabesques, skaters play with their average distance from the axis of rotation. They slow down when they

Ice skating spin. By pulling in her arms, and then raising them above her head, the skater decreases her moment of inertia. In the left position with arms spread, $I{=}1.9\, kg \cdot m^2$ and the rotational velocity is 2 full rotations per second, while on the right position, $I{=}0.8\, kg \cdot m^2$ and the rotational velocity is 4.75 full rotations per second, which is very fast !

spread their arms or legs, and speed up when the body is bunched up.

WHY IS THE ROTATION OF THE BODY IN SOMERSAULTS SO FAST?

Depending on their position, divers, gymnasts who have let go of the bar, or trampoline jumpers can turn faster or slower. The more they are grouped together in the somersault, the faster they turn; the more they extend their body, the slower they turn. The moment of inertia depends on the distribution of mass according to distance from the axis of rotation. There are several possible axes of rotation for a human body: the vertical axis is used for ice-skating spins or diving twists, the horizontal axis transverse to the waist for somersaults, and the horizontal axis perpendicular to the body for wheels.

A gymnast springs from a bar. Parabolic movement of the centre of gravity and rotation around the centre of gravity. Grouped position in the somersault. You slow down by unfolding your arms and legs.

In fact, the diver or gymnast who springs from the bar combines two movements:

- The movement of the centre of gravity, determined by weight and initial propulsion speed. This is a vertical free-fall movement in the case of the diver, and a parabola in the case of the gymnast.
- Rotation around the centre of gravity. As the weight passes through the centre of gravity, the angular momentum is preserved. You accelerate by grouping the body and slow down by unfolding arms and legs.

In somersaults, angular momentum (moment of inertia multiplied by rotation speed) is conserved. Increasing the moment of inertia reduces the speed of rotation, and vice versa. In the **Moment of inertia** box, the table of moment of inertia values shows that in the grouped somersault position, the moment of inertia is about 5 times smaller than in the extended body position. The speed of rotation is therefore multiplied by 5. At the finish, the gymnast has to lock her position at a low angular velocity. With practice, the athlete finds the exact moment to straighten up. If she is too late, she falls on her back, too early and her feet touch the ground at too great a speed and she falls down.

CHAPTER 6

COUNTING

WHY SHOULDN'T YOU RELY ON DOPING TESTS TOO MUCH?

The anti-doping agency uses tests to determine the presence of illegal substances in athletes. And yet, a small calculation of probability shows that tests that are considered reliable can produce a much higher number of false positives than expected. This can have dramatic consequences for athletes.

The sensitivity of a test determines a threshold at which athletes are considered positive. How should this threshold be set? A high threshold increases the number of false negatives (the test is negative, even though doping has taken place) and reduces the number of false positives (the test is positive, even though doping has not taken place) because the substance may be produced naturally by the body. On the other hand, a low threshold reduces the number of false negatives but increases the number of false positives. The choice of the threshold can therefore have significant consequences.

Let us imagine that the reliability rate of a test is 95%: out of 100 tests carried out, 95 will give correct results and 5 false

© The Author(s), under exclusive license to Springer Nature Switzerland AG 2024

A. Aftalion, *Be a Champion*, Copernicus Books,
https://doi.org/10.1007/978-3-031-54082-0_6

results. In other words, if an athlete is doping, there is a 0.95 probability that the test will be positive, and a 0.05 probability that it will be negative.

Let us look at the problem the other way round: if a test is positive, what is the risk of falsely accusing a non-doped athlete? To answer this question, we need to make a tree of all the possibilities. Let us imagine that 10% of athletes are doped and 90 % are not. Then on N tests, the number of positive tests is equal to the sum of those carried out on doped athletes who tested positive (whose tests were accurate) and those carried out on non-doped athletes who tested positive, but whose tests were false. These are the two red paths on the diagram below. This provides $(0.10 \times 0.95 + 0.90 \times 0.05)N$ positive tests.

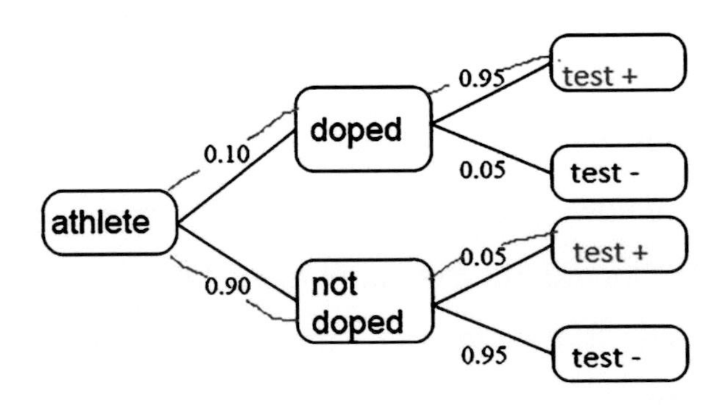

Probability tree diagram describing positive and negative tests.

If a test is positive, the probability that it concerns an athlete who is not doping is equal to the number of positive tests of a non-doped athlete out of the total number of positive tests, i.e.,

$$\frac{(0.90 \times 0.05)N}{0.10 \times 0.95 + 0.90 \times 0.05)N} = 0.32.$$

Therefore, there is almost a one in three chance that an athlete who is not doping will nevertheless test positive, which is huge, especially given the high reliability rate of the test. And this can have dramatic consequences for an athlete's career!

HOW DOES ARTIFICIAL INTELLIGENCE HELP SOCCER COACHES?

The introduction of data analysis and statistics in sports began with baseball, basketball, and tennis. In soccer, the use is more recent. There are a lot of players, they move around a lot, but in the end very few goals are scored. So it is not so easy to characterise the decisive phases of the game, because one can't just link actions and goals.

Artificial intelligence is based on three coupled approaches: statistical learning, computer vision, and game theory.

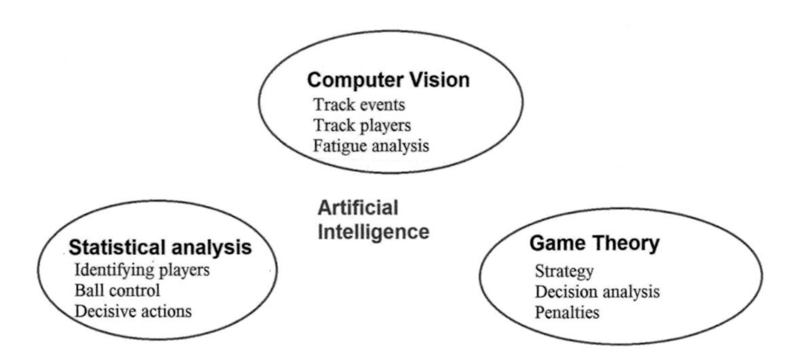

Diagram summarising artificial intelligence for soccer coaches.

Game theory is the mathematical study of the strategies used by players to identify trends and predict certain outcomes. It applies particularly well to penalties, for example. The player who takes one penalty is faced with a binary choice: left or right. The goalkeeper in front of him can also choose to dive to one side or the other. In this type of game, you only win if the other player loses; it is called a zero-sum game. Each shooter is more accurate on one side than the other. Each goalkeeper adjusts his behaviour according to the shooter's preferences and the history of penalties he has taken in the past. Thanks to probability, game theory, and the notion of Nash equilibrium, wc can analyse things more precisely and gain a certain amount of information. For example, game theory shows that, for both the player and the goalkeeper, it is more effective to choose the side at random, so as not to give any usable indication.

Nowadays, artificial intelligence can be used to detect, recognise and track the ball or a player in real time from a filmed match. Some models, as well as identifying players with a high degree of accuracy, can detect events such as goals, passes, and fouls. At the end of a match, we therefore have precise, quantitative information on each player's fouls, his actions according to his position on the pitch, or his actions that led to decisive events. Video coupled with artificial intelligence now helps coaches to extract relevant information and isolate explicit sequences of play, so that they can better pinpoint strengths and tactics, or on the contrary, weaknesses and mistakes. This makes it possible to improve a team's training, but also to study an opponent's strategy or weak points and prepare a winning game strategy.

These models can also be used to extract statistics on players' positions on the pitch. By combining positional and event data, it is possible to obtain relevant, summarised information. For example, the number of successful passes can be calculated to assess a player's performance in a particular sector of the game and determine the best position for him in the team. A circle of influence is associated with each player in the algorithm. The larger the radius, the greater the influence; the darker the radius, the more dangerous the actions. Algorithms can also predict the best selection of players for a national team.

To help recruiters in their decision-making, data analysis can identify players' qualities and weaknesses quantitatively and accurately. Some clubs are now able to algorithmically simulate matches with the players they are targeting to better ensure their compatibility with the team and decide on the best place to give a player.

It is not enough to collect a large amount of data, you also have to be able to analyse it to make a prediction. This comes up against the psychological dimension, but we are beginning to be able to model this too. And what if algorithms were to replace coaches?

WHAT IS THE FASTEST SPORT?

It all depends on how you understand the question.

In 2019, Eliud Kipchoge became the first able-bodied runner to complete a marathon in under two hours. But wheelchair racers have been breaking this barrier for decades. The fastest official wheelchair time was set in 1999 by Switzerland's Heinz Frei, 40 minutes quicker than Kipchoge. In fact, on all Olympic distances of at least 800 m, the fastest wheelchair athletes outstrip the able-bodied: from 800 m upwards, whatever the distance, the speed of the wheelchair athletes holding the records is almost constant (8.68 m/s on an 800 m, 8.54 on a 1,500 m, 8.16 on a 10,000 m and 8.77 on a marathon). Why are marathon runners faster? Because a marathon is held on roads rather than on an athletics track, and the looser bends are easier for wheelchairs to negotiate. Over short distances, wheelchair records are obviously affected by a slower start than that of a running athlete. The small variation in speed over distances also enables para-athletes to take part in more races of different lengths. Julien Casoli, who has won the Paris marathon several times in the disabled sport category, has also won medals at European and world championships in the 800 m, 1,500 m, and 5,000 m events.

Which sports can achieve the highest speed? The record is held by the badminton shuttlecock, at 351 mph, while the golf ball can reach 241 mph and jai alai 194 mph. If we are talking about the sportsman himself, the record is held by a speed skier, with 158 mph reached in less than 5 seconds, i.e., faster acceleration than a Formula 1 car.

The question also arises as to which sport requires the fastest reaction speed. The fastest tennis serves reach a speed of around 160 mph. If we imagine that the opponent is directly diagonally across the court, the ball has to travel around 25 m, leaving 34 milliseconds for the player to react and catch the ball. As a result, the best players' serves are often winning (known as "ace" serves). In baseball, the time between pitch and strike is estimated at 40 milliseconds, during which the batter, 18 m from the bat, must locate the ball, estimate its trajectory, decide whether or not to swing, and finally strike in the right spot.

WHY DOES THE DECATHLON SCORING SYSTEM FAVOUR SPRINTERS?

World record-holder since 2018 and holder of numerous medals at the Olympic Games and world championships, Kevin Mayer, has been making France shine in the decathlon event for several years now. The decathlon consists of 10 track and field events spread over two days, including races (100 m, 400 m, 1,500 m, 110 m hurdles), jumps (long jump, high jump, pole vault) and throws (shot put, discus, javelin). In order to combine the results of these events — some provide times and some give distances — a point system has been developed. Each performance is awarded a predetermined number of points according to a set of performance tables. These are added, event by event, and the winner is the athlete with the highest points total after ten events.

Three types of events: race, jump, throws.

The scoring table takes the form of a thick book, since for each of the ten events, the progression of performance is shown from point to point, from 1 to around 1,200. The minimum performance is at a very low level (well below regional level), the maximum performance above the world record for the event. If you were able to break the world record in every event, then you would score over 12,000 points. In fact, there is a formula for calculating event points based on performance p (time in seconds for races, distance in cm for jumps and m for throws). The formula is as follows:

$$f(p) = a\,|b - p|^c,$$

where a, b and c are fixed parameters which depend on the event:

Event	*a*	*b*	*c*
100 m	25.4347	18	1.81
Long jump	0.14354	220	1.4
Shot put	51.39	1.5	1.05
High jump	0.8465	75	1.42
400 m	1.53775	82	1.81
110 m hurdles	5.74352	28.5	1.92
Discus throw	12.91	4	1.1
Pole vault	0.2797	100	1.35
Javelin throw	10.14	7	1.08
1 500 m	0.03768	480	1.85

Coefficients used to score points in a decathlon event.

We can then calculate how to get 900 points per event, which leads to a total close to Kevin Mayer's world record of 9,126 points (see the table below). Note that Kevin Mayer is better in the sprint and long jump than in shot put, discus throw, high jump or 1,500 m. This is true of all decathlon champions, as can be seen by analyzing the average performances per event of the top 100 (last column in the table):

Event	900 pts/event	9 126 pts (Kevin Mayer)	Mean for the top 100 athletes
100m	10.83 s	10.55 s (963 pts)	10.76 s (915 pts)
Long jump	7.36 m	7.8 m (1010 pts)	7.66 m (975 pts)
Shot put	16.79 m	16.00 m (851 pts)	15.47 m (819 pts)
High jump	2.10 m	2.05 m (850 pts)	2.06 m (859 pts)
400 m	48.19 s	48.42 s (889 pts)	48.22 s (899 pts)
110 m hurdles	14.59 s	13.75 s (1007 pts)	14.23 s (945 pts)
Discus throw	51.4 m	50.54 m (882 pts)	46.92 m (807 pts)
Pole vault	4.96 m	5.45 m (1051 pts)	4.95 m (895 pts)
Javelin throw	70.67 m	71.90 m (918 pts)	64.46 m (832 pts)
1,500 m	4 mn 7.4 s	4 mn 36.11 s (705 pts)	4 mn 34.12 s (718 pts)

Performance in decathlon per event. Column 1, to get 900 points. Column 2, Kevin Mayer's points to get 9126 points. Column 3, mean of the top 100 athletes.

Averaged over the top 100 athletes, performances are excellent in sprinting (100 m and 400 m), 110 m hurdles and long jump, all of which are based on short, fast running. On the other hand, they are very poor in the 1,500 m and throwing events (shot put, discus throw, javelin). This means that the athletes who take up the decathlon are sprinters, not throwers or endurance athletes, which has a lot to do with the way points are counted, and does not favor throws.

Event	$pf'(p)/(p)$ $=cp/\|p\text{-}b\|$	Points obtained by an improvement of 5%
100 m	2.73	122.9
Long jump	2.0	90.0
Shot put	1.15	51.75
High jump	2.20	99.0
400 m	2.57	115.67
110 m hurdles	2.01	90.5
Discus throw	1.19	53.5
Pole vault	1.69	76.0
Javelin throw	1.19	53.5
1 500 m	1.96	88.2

Points obtained by improving performance by 5%.

Indeed, if we imagine a 5% improvement in performance over the one required to obtain 900 points, we can calculate the number of points this would earn in the next table. The calculation of the points obtained is given by the formula

$$p\frac{f'(p)}{f(p)} \times 900 \times 0.05.$$

Note that the number of points earned is very high for sprints and jumps (100 points or more) and very low for throws (50 points). So it is more rewarding to be better than average in sprints than in throws.

You might ask what would happen if you changed the points table. If we go back to the coefficients in the very first table, we notice that b is a threshold value which does not bring any points, and that the better the performance, the greater $|p-b|$ becomes. As c is greater than 1, the gain in points from improved performance is greater for good than for poor performance. On the other hand, the coefficient c is around 1.8 for running, 1.4 for jumping and 1.05 for throwing, so the gain in points is greater when improving a running performance than a throwing one.

WHY DOES THE SCORING METHOD IN TENNIS LENGTHEN THE GAME?

Anyone who follows a tennis match may be surprised by the way points are scored. To win a match, you need to win two sets for women or three sets for men. Sets are played over 6 games, provided there is a two-game difference between the players. To win a game, you must score at least 4 points, and, after the first 3, be 2 points clear of your opponent. Points are not counted from 1 to 4, but are worth 15 for the first, 30 for the second and 40 for the third. Beyond that, and if there is no two-point difference, the player who scores a point gets an ad-in or ad-out depending on whether he is serving or receiving (ad meaning advantage).

It would be simpler to count the points 1 by 1 in a cumulative way and announce a score, as in handball or table tennis.

Having to win a few points to win a game, then a few games to win a set, then a few sets to win a match, is to the advantage of the better of the two players.

What is the probability of winning a game when the players are of unequal strength, and how does this probability change over the course of the game? For simplicity's sake, we will assume that the relative strengths between two opponents are measured by the percentage each has of winning a point against the other. This is not exactly the case, as these percentages vary, in practice, with the stamina of each player, the position of serving or not, and so on.

Why are tennis points worth 15 for the first, 30 for the second and 40 for the third?

This particular count comes from jeu de paume, invented by French monks and widely popularized during the Renaissance. It was played on a rectangular court separated into two sections of 60 feet each by a net. Parallel to the net were lines numbered 15, 30 and 40, depending on the distance in feet from the bottom of the court. Once a point was won, the player who had won had to move to the next line, bringing him a little closer to the net.

The word "tennis" itself comes from the old French "tenez", pronounced at jeu de paume before the serve to warn your opponent that you are about to serve. It was then deformed into English to become "tenetz", "tenes" and finally "tennis". English tennis terms also come from old French: "deuce" comes from the French deux to indicate that there are two points left to be won; "love" designates zero and comes from the word "l'œuf" (the egg in French), which is indeed shaped like a zero!

For example, if we estimate that the probability of X winning a point against Y is $p=0.6$, this means that, on average, X wins 6 points out of 10 against Y, while Y wins only 4 out of 10 against X (probability $1-p=0.4$ or, in percentage terms, 40%). This may, for example, be the result of a statistic evaluated on games in which the two players have previously played against each other.

We can then calculate the probability that a player X wins the game against a player Y, depending on the score. In the table below (see the box **Calculating your chances of winning** for details on how to fill it in), we see that the probability that X wins the game when the score is 0-0 is 73%, almost 3 out of 4 chances, much greater than the 6 out of 10 chances he has of winning a point.

Table of probabilities that X wins against Y depending on the score, knowing that

probability of X winning against Y 60%

probability of Y winning against X 40%

Y\X	0	15	30	40	Game
0	73.57%	84.21%	92.71%	98.03%	100.00%
15	57.62%	71.45%	84.74%	95.08%	100.00%
30	36.89%	51.51%	69.23%	87.69%	100.00%
40	14.95%	24.92%	41.54%	69.23%	
Game	0.00%	0.00%	0.00%		

If the score is 40-0 for X, then he has 98 chances out of 100 of winning the game, and about 7 out of 10 if the players are deuce 40-40.

Other example:

Table of probabilities that X wins against Y depending on the score, knowing that

probability of X winning against Y 75%

probability of Y winning against X 25%

Y\X	0	15	30	40	Game
0	94.92%	97.56%	99.14%	99.84%	100.00%
15	87.01%	92.81%	97.03%	99.38%	100.00%
30	69.61%	80.16%	90.00%	97.50%	100.00%
40	37.97%	50.63%	67.50%	90.00%	
Game	0.00%	0.00%	0.00%		

Here, in the case where the probability of X winning against Y is 75% (i.e., 3 times out of 4), X has, at the start of the game, a 95% chance of winning, and has 9 chances out of 10 if the players are "tied" at deuce (40-40).

Unlike a simple cumulative score (21-18, for example), the successive counting of points, then games, then sets, exaggerates the differences in strength between the players, but maintains the suspense much more! Indeed, the interest of the match is maintained for a longer time, since with each game, then each set, the score falls to zero and the competition is restarted. Whereas if, after losing a set 6-0, a player had to continue with a cumulative score, spectators' interest would be diminished, as the player would have to come back up with all the lost points, and the match would end more quickly.

Calculating your chances of winning

The calculation can easily be carried out using an Excel spreadsheet.

Suppose you want to evaluate the probability $P_{a,b}$ of winning the game when the score is (a,b), for the player whose probability of winning each point is p (with $q = p\text{-}1$).

To win, the player must:

- either win the next point (this probability is p) and then win the game, the score then being (a+1,b), (this probability is $P_{a+1,b}$)

- or lose the next point (this probability is q) then win the game, the score then being (a,b+1), (this probability is $P_{a,b+1}$).

So it is easy to see: $P_{a,b} = p\ P_{a+1,b} + q\ P_{a,b+1}$. Note that the calculation of the odds of winning uses two elementary rules of probability calculus:

-to calculate the probability that (either A occurs or B occurs), we add the probabilities that (A occurs) and that (B occurs).

-to calculate the probability that (A occurs, then B occurs), multiply the probabilities that (A occurs) then that (B occurs, knowing that A has occurred).

The probability in the box in the excel table is therefore equal to p times that of the box on the right plus q times that of the box below, as indicated by the arrows in the table.

All that remains to complete the excel table is to fill in the right-most and bottom-most squares, so that the table can be completed, square by square, up to square (0,0):

	0	15	30	40	Game
0					100.00%
15		+ ←			100.00%
30		↑			100.00%
40			↑ + ←	E	
Game	0.00%	0.00%	0.00%		

The cells in column 4 are 100%, while those in row 4 are 0. We still need to calculate the probability (E) of winning in the deuce case.

We are looking for the probability of winning the game, starting from a deuce, for the player who has a probability of winning each point equal to p.

To win, he must :

- either win the next point (probability p) and then win the next point again (probability p),

- or lose the next point (probability $1-p=q$) and then win the next point (otherwise he has lost the game), leaving him with probability E of winning the game.

- or, conversely, win the next point and lose the one after that, ending up with probability E of winning the game.

Therefore, we have: $E = p^2 + 2\ q\ p\ E$, since probabilities add up for events of the either-or type, and multiply for successive events. The calculations yield $E(1-2qp) = p^2$. So, in the case of $p=0.6$, we have $E = 0.6923$ (box at bottom right of table).

Note that if p is close to ½, for example $p = 1/2+\varepsilon$, where is ε small, then we can calculate the probability of the player of winning the game which is at leading order $p = 1/2+5\varepsilon/2$.

The probability is therefore greater of winning a game than of winning a point. We see that for $\varepsilon = 0.1$, this gives a probability of winning a game of 0.7, not far from the 0.73 in the table. The probability of winning a match in three sets becomes $½+9\varepsilon/2$.

It has increased again!

IS THERE A LAW OF EVOLUTION OF RECORDS?

Sports records are regularly broken. It is their destiny! But how? Is there a general law? Is there a human limit?

Three phenomena in general lead to improvements in performance: technological advances in equipment, technical innovations, or the emergence of a champion or coach who stimulates others.

The most obvious technological advances are in materials for javelins or poles, and wetsuits in swimming. Technical innovations include new postures such as those of Dick Fosbury in the high jump or Bob Beamon in the long jump, and improvements in swimming techniques.

Depending on the discipline, for some records one seems to be close to the human limit, while for others we can hope that improvements are still possible. Let us take a closer look at some sports records.

High Jump (men)

The record for the men's high jump has remained unchanged since 1993, when Javier Sotomayor broke the 2 m 45 mark. One might think that the limit in this discipline has been reached, or at any rate that there will be very little further progress. What about the history of the high jump record?

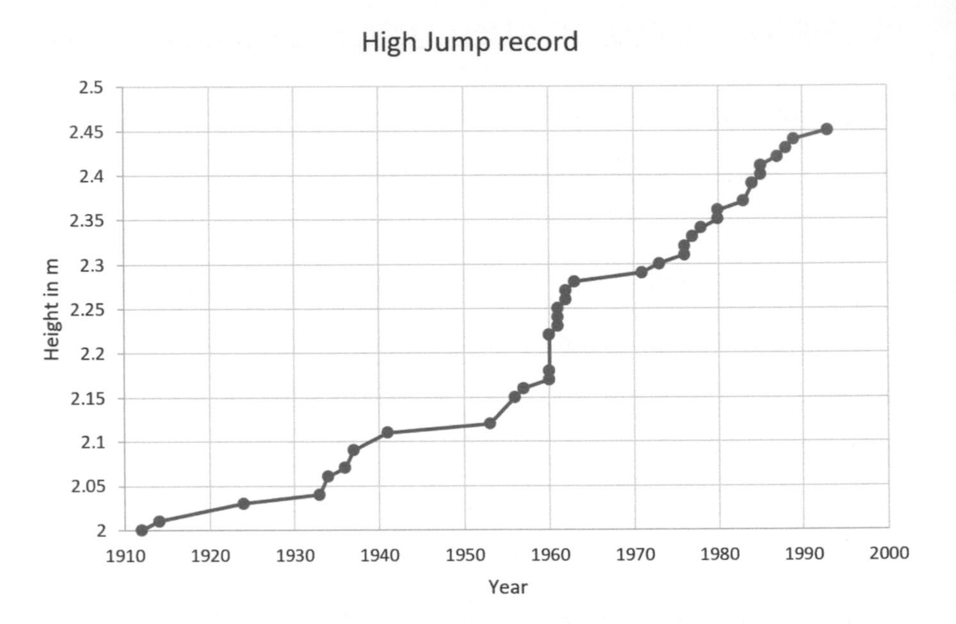

Before 1900, athletes jumped high in a scissor jump, bringing both legs up and jumping in a virtually seated position. Then it was discovered that it was possible to jump higher by leaning the chest backwards until you were lying on your side. This was the western roll technique, adopted, for example, by Harold Osborn, world champion in 1924 with a height of 2.03 metres.

In 1936, a new technique appeared where the athlete cleared the bar face-down, with the body stretched along the bar; the straddle technique. The straddle allowed parts of the legs to be lower than the bar at the peak of the jump. John Thomas and then Valery Brumel improved the straddle into the dive straddle and were the great champions of the 1960s. They raised the record to 2.28 m. Between 1961 and 1963, Valery Brumel beat his world record six times, improving it by 1 cm each time.

The appearance of these two new techniques in 1936 and 1960-63 corresponded to very significant changes in the slope of the record curve.

It was then that Dick Fosbury, at the 1968 Olympic Games in Mexico City, introduced a spectacular new technique, which consisted of standing with his back to the vault! By tilting his shoulders backwards as he rose, he raised his pelvis much higher than with the belly roll, and he could also let his legs hang behind him for a moment, unfolding them only to let them pass the bar.

This technique became known as the Fosbury flop. In Mexico City, Dick Fosbury cleared his first 5 bars, from 2.03 m to 2.20 m, on the first attempt, to end up with his compatriot Caruthes at 2.22 m. Then he was the only one to pass 2.24 m, and became Olympic champion. The judges initially rejected his jumps, but then accepted them after checking that the only requirement in the rules was that he take off with just one foot.

It took some time for the champions to adopt this technique. The era of the Fosbury flop began in earnest in 1973, with a record of 2.30 m, until 1993, when Javier Sotomayor broke the 2.45 m barrier. It is 30 years since that record was set! However, from 1973 to 1993, the record's progression curve was almost a straight line, which might have suggested that it could still be beaten.

So it is the advent of technical progress that suddenly changes the situation, and the record curve jumps up and down, without giving any indication of the future.

800 m (men)

The record for the 800 metres in athletics was 1'51"9 in 1912, when the world federation decided to approve the main world records; today it is 1'41"91, so ten seconds below.

To discuss this record since 1972, it is interesting to replace the points representing the records by a curve that passes as closely as possible to the points. This is called interpolation.

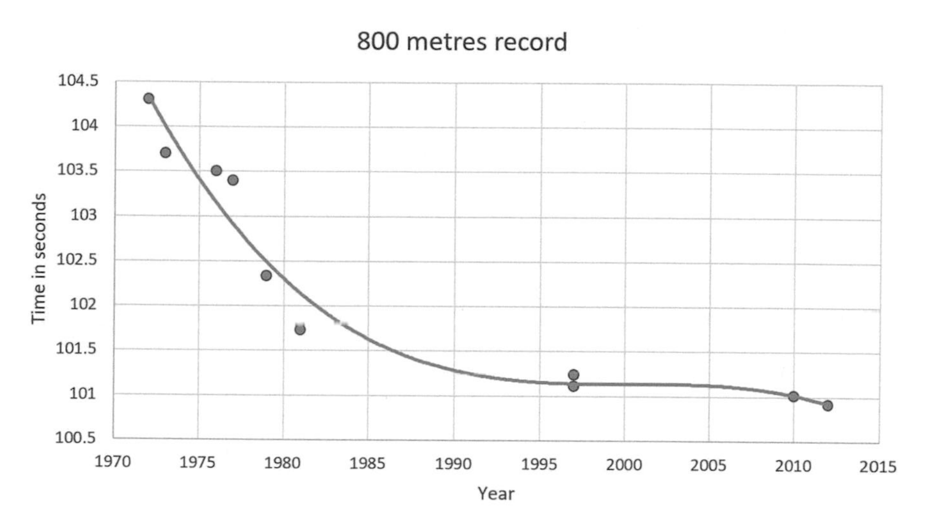

The record decreases, but less and less until 1997. This is reflected by the fact that the tangent to the red curve has a negative slope, but that the slope decreases in absolute value. It thus seems to become almost horizontal, with the ordinates of successive points tending towards a limit which would be that of human possibilities. In this way, the curve representing the record measurement as a function of time would present what mathematicians call an asymptote.

But in 2010, David Rudisha beat the record held by Wilson Kipketer by 2/100ths of a second and then by 1/10th. At the 2012 London Olympics, David Rudisha, in a race in which he was in the lead from start to finish, beat his own record by going under the 101 s mark, i.e., under 1 minute 41 seconds. He has been unbeaten ever since. Once again, an extraordinary champion has changed the direction of the curve, and it is hard to know which way it will go from now on.

Swimming, 100 m free style

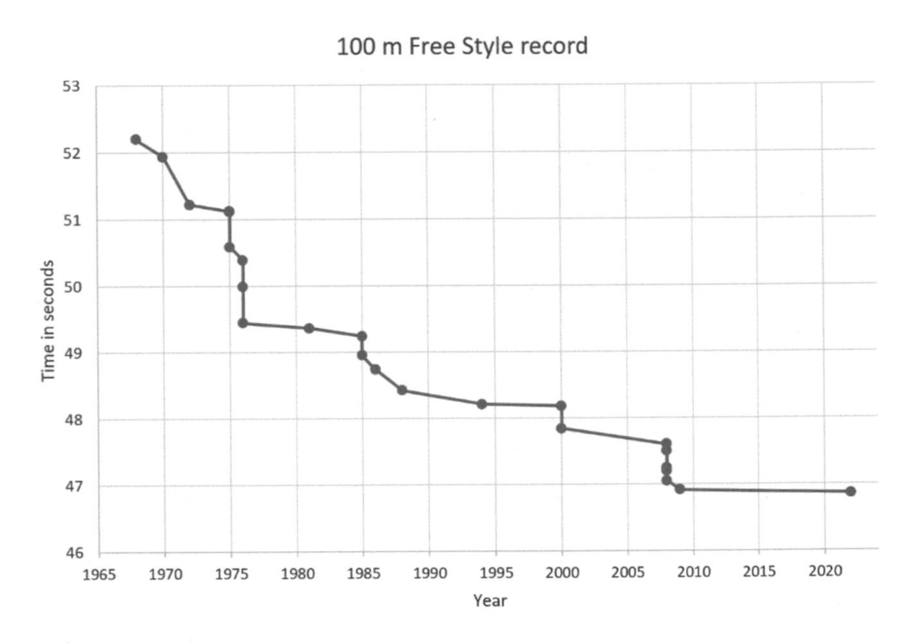

The 100 m freestyle record curve as a function of time shows a gradual decline. In some years, there are vertical sections, meaning that the record has been broken several times, while in other periods the change is slower. Some vertical or almost vertical sections of the curve correspond to extraordinary champions: Mark Spitz in 70-72, Jim Montgomery in 75-76, Matt Biondi in 85-88, Alain Bernard and Eamon Sullivan in 2008. The champions of the 1970s would not be selected for the Olympic Games today.

Two years stand out in particular: 1976 and 2008. The first corresponds to an evolution in swimming technique developed by the American coach James Counsilman. Then

2008 was a key year, with the introduction of wetsuits and the increase in pool depth from 2 to 3 metres, which reduced swimming resistance. After the ban on wetsuits, it took some time for the record to be broken again, but it happened in 2022! It is hard to predict where this curve will lead.

100 metres

Since 2009, Jamaican Usain Bolt has held the world record for the 100 metres in 9.58 seconds, set in Berlin at the World Championships. His record has been unbeaten ever since. In Berlin in 2009, Bolt beat his own personal best, set at the Beijing Olympics the previous year, by 0.11 seconds. But it should have taken 14 years if we had imagined extending the curve.

The curve showing the 100 metres record time over the years is incredible. The concavity of the curve is the opposite of what you would expect and, what is more surprising, the unlikely asymptote resembles a vertical rather than a horizontal line. The almost vertical parts correspond to extraordinary champions: Carl Lewis from 1987 to 1991 and Usain Bolt for the latest records from 2007 to 2009.

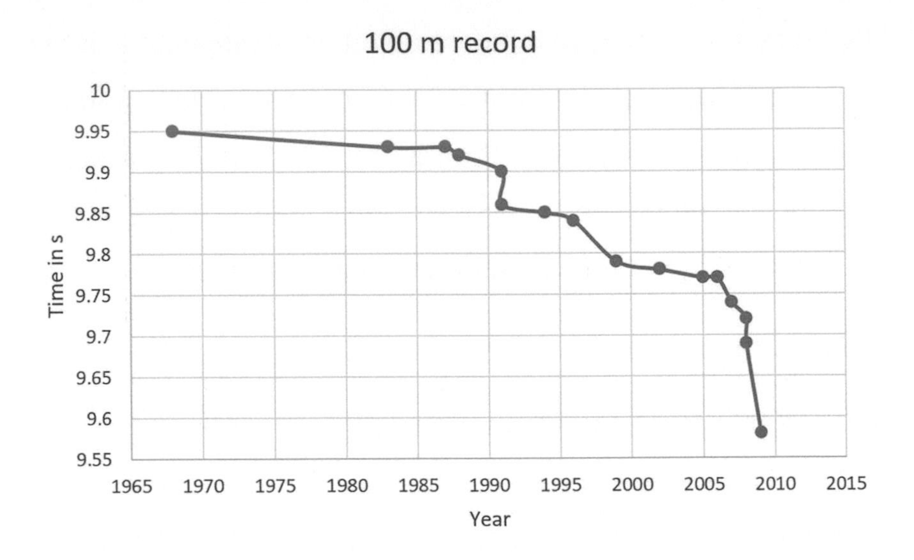

This shows the importance of exceptional individuals in the world of champions … and the difficulty of fitting them into a model!

There has been much speculation about the possible limit of the 100 metres record. Forty years ago, it was predicted that the human record would stand at 9.60 seconds over 100 metres. Bolt did better! Since then, people have wondered whether we could reach 9.55, 9.48, 9.29 seconds. For his part, Bolt believes that the record will not fall below 9.40 seconds.

Before him, probabilistic models indicated that the probability of breaking the 9.58 seconds barrier was 0.64%, proof that Bolt is indeed a phenomenon! This changes our perceptions of human ability. Today, the probability of passing the 9.55-second mark is estimated at 1.02% and that of going under 9.5 at 0.52%.

Photo by Juice Dash/Shutterstock.com

Numerous mathematical models have attempted to establish a law of record evolution. In a number of sports, there is an exponential decrease and the time to break a record is becoming longer and longer. Some estimates say that by 2027, records will not have improved by more than 0.05% if conditions do not change. But technology and technique continue to evolve, and so do the rules.

To this day, humans remain unpredictable and extraordinary athletes continue to impress us. Michael Phelps, the American swimmer who broke 37 world records, is quoted as saying: "You can't put a limit on anything. The more you dream, the further you go!"

FURTHER READING

Chapter 1

This chapter is based on Amandine Aftalion's research.

Aftalion, A. (2017). How to run 100 meters. *SIAM Journal on Applied Mathematics*, *77*(4), 1320-1334.

Aftalion, A. & Martinon, P. (2019). Optimizing running a race on a curved track. *PloS One*, *14*(9), 0221572.

Aftalion, A. & Trélat, E. (2020). How to build a new athletic track to break records. *Royal Society Open Science*, *7*(3), 200007.

Aftalion, A. & Trélat, E. (2021). Pace and motor control optimization for a runner. *Journal of Mathematical Biology*, *83*(1), 9.

Mercier, Q., Aftalion, A. & Hanley, B. (2021). A model for world-class 10,000 m running performances: Strategy and optimization. *Frontiers in Sports and Active Living*, 226.

Wilson, D., Bryborn, R., Guy, A., Katz, D., Matrahazi, I., Meinel, K., Salcedo J., Wauhkonen, K. (eds) (2008), IAAF track and field facilities manual. Monaco: Editions EGC.

© The Author(s), under exclusive license to Springer Nature Switzerland AG 2024
A. Aftalion, *Be a Champion*, Copernicus Books,
https://doi.org/10.1007/978-3-031-54082-0

Chapter 2

Griffing, D.F. (1995). *The Dynamics of Sports—Why That's the Way the Ball Bounces*. Kendall Hunt Pub Co.

Hay, J. (1978). *The biomechanics of sports techniques*. Prentice-Hall.

Helene, O. & Yamashita, M. T. (2005). A unified model for the long and high jump. *American journal of physics*, *73*(10), 906-908.

Herman, I.P. (2016). *Physics of the human body*. Springer.

Lichtenberg, D. B. & Wills, J. G. (1978). Maximizing the range of the shot put. *American Journal of Physics*, *46*(5), 546-549.

Lithio, D. & Webb, E. (2006). Optimizing a volleyball serve. *Rose-Hulman Undergraduate Mathematics Journal*, *7*(2), 11.

McGinnis, P.M. (1997). Mechanics of the pole vault take-off. *New studies in athletics*, *12*, 43-46.

Tan, A. & Zumerchik, J. (2000). Kinematics of the long jump. *The Physics Teacher*, *38*(3), 147-149.

White, C. (2010). *Projectile dynamics in sport: principles and applications*. Routledge.

Chapter 3

Alam, F., Steiner, T., Chowdhury, H., Moria, H., Khan, I., Aldawi, F. & Subic, A. (2011). A study of golf ball aerodynamic drag. *Procedia Engineering*, *13*, 226-231.

Barton, N.G. (1982). On the swing of a cricket ball in flight. *Proceedings of the Royal Society of London. A. Mathematical and Physical Sciences*, *379*(1776), 109-131.

Bower, R. & Cross, R. (2005). String tension effects on tennis ball rebound speed and accuracy during playing conditions. *Journal of sports sciences*, *23*(7), 765-771.

Brancazio, P.J. (1981). Physics of basketball. *American Journal of Physics*, *49*(4), 356-365.

Brancazio, P. J. (1985). The physics of kicking a football. *The Physics Teacher*, *23*(7), 403-407.

Brody, H. (1979). Physics of the tennis racket. *American Journal of physics*, *47*(6), 482-487.

Brody, H. (1981). Physics of the tennis racket II: The "sweet spot". *American Journal of Physics*, *49*(9), 816-819.

Brody, H. (1997). The physics of tennis. III. The ball–racket interaction. *American Journal of Physics*, *65*(10), 981-987.

Briggs, L.J. (1959). Effect of spin and speed on the lateral deflection (curve) of a baseball; and the Magnus effect for smooth spheres. *American Journal of Physics*, *27*(8), 589-596.

Carré, M. J., Asai, T., Akatsuka, T. & Haake, S. J. (2002). The curve kick of a football II: flight through the air. *Sports Engineering*, *5*(4), 193-200.

Chowdhury, H., Loganathan, B., Wang, Y., Mustary, I. & Alam, F. (2016). A study of dimple characteristics on golf ball drag. *Procedia engineering*, *147*, 87-91.

Clanet, C. (2015). Sports ballistics. *Annual Review of Fluid Mechanics*, *47*, 455-478.

Cross, R. (1998). The sweet spots of a tennis racquet. *Sports Engineering*, *1*(2), 63-78.

Erlichson, H. (1983). Maximum projectile range with drag and lift, with particular application to golf. *American Journal of Physics*, *51*(4), 357-362.

Mehta, R., Alam, F. & Subic, A. (2008). Review of tennis ball aerodynamics. *Sports technology*, *1*(1), 7-16.

Watts, R.G. & Sawyer, E. (1975). Aerodynamics of a knuckleball. *American Journal of Physics*, *43*(11), 960-963.

Wesson, J. (2019). *The science of soccer*. CRC Press.

Chapter 4

Anderson, B.D. (2008). The physics of sailing. *Physics Today, 61*(2), 38.

Armenti Jr., A. (1984). How can a downhill skier move faster than a sky diver? *The Physics Teacher, 22*(2), 109-111.

Armenti, A. (1985). Why is it harder to paddle a canoe in shallow water? *The Physics Teacher, 23*(5), 310-313.

Barbosa, T.M., Marinho, D. A., Costa, M. J. & Silva, A.J. (2011). Biomechanics of competitive swimming strokes. *Biomechanics in applications*, 367-388.

Herreshoff, H.C. & Newman, J. N. (1966). The study of sailing yachts. *Scientific American, 215*(2), 60-71.

Hignell, R. & Terry, C. (1985). Why do downhill racers pre-jump? *The Physics Teacher, 23*(8), 487-487.

Landell-Mills, M.N. (2022). How fish swim according to Newtonian physics. Researchgate preprint.

Parolini, N. & Quarteroni, A. (2005). Mathematical models and numerical simulations for the America's Cup. *Computer Methods in Applied Mechanics and Engineering, 194*(9-11), 1001-1026.

Quarteroni, A. (2022). Mathematics in the Wind. In *Modeling Reality with Mathematics* (pp. 67-84). Cham: Springer International Publishing.

Toussaint, H.M., Hollander, A.P., Van den Berg, C. & Vorontsov, A. (2000). Biomechanics of swimming. *Exercise and sport science*, 639-660.

Wilson, D.G., & Schmidt, T. (2020). *Bicycling science*. MIT press.

Chapter 5

Braune, W. & Fischer, O. (2013). *Determination of the moments of inertia of the human body and its limbs*. Springer Science & Business Media.

Frohlich, C. (1979). Do springboard divers violate angular momentum conservation? *American journal of physics*, *47*(7), 583-592.

Frohlich, C. (1980). The physics of somersaulting and twisting. *Scientific American*, *242*(3), 154-165.

Jones, D. E. (1970). The stability of the bicycle. *Physics today*, *23*(4), 34-40.

Lowell, J. & McKell, H.D. (1982). The stability of bicycles. *American Journal of Physics*, *50*(12), 1106-1112.

Smith, T. (1982). *Gymnastics: A mechanical understanding.* Holmes & Meier Pub.

Chapter 6

Barrow, J.D. (2013). *Mathletics,* a scientist explains 100 amazing things about sports. Random House.

Chang, Y.S., & Baek, S.J. (2011). Limit to improvement in running and swimming. *International Journal of Applied Management Science*, *3*(1), 97-120.

Fischer, G. (1980). Exercise in probability and statistics, or the probability of winning at tennis. *American Journal of Physics*, *48*(1), 14-19.

Newton, P.K. & Keller, J.B. (2005). Probability of winning at tennis I. Theory and data. *Studies in Applied Mathematics*, *114*(3), 241-269.

Noubary, R. (2010). What is the Speed Limit for Men's 100 Meter Dash. *Mathematics and Sports*, *43*, 287.

Tuyls, K., Omidshafiei, S., Muller, P., Wang, Z., Connor, J., Hennes, D. et al. (2021). Game Plan: What AI can do for Football, and What Football can do for AI. *Journal of Artificial Intelligence Research*, *71*, 41-88.

World records are available on Wikipedia.

ACKNOWLEDGEMENTS

This book originated from joint works with Emmanuel Trélat on the mathematical modeling of running. It was an honour and a pleasure to work with him and I would like to thank him warmly for his constant enthusiasm and support.

I am also extremely grateful to André Deledicq, since this book was born after many discussions together on how to make mathematics and sports popular.

A lot of colleagues helped me with their remarks, suggestions, support, enthusiasm: Paul Martins, Pierre-Michel Menger, Qiang Du, Michèle Leduc, Michel Le Bellac, Etienne Sandier, Raphael Danchin, Charles-Edouard Le Villain. I would like to thank them all here. Working on interdisciplinary topics, on the edge between mathematics and physics, is due to the influence of many professors, and in particular Etienne Guyon whom i would like to pay tribute to here.

I am very grateful to my Editor, Elizabeth Loew, as well as the to the whole team at Springer.

And last but not least I want to thank my daughter Hélène for her very careful reading of the manuscript and her smart suggestions.

© The Author(s), under exclusive license to Springer Nature Switzerland AG 2024
A. Aftalion, *Be a Champion*, Copernicus Books,
https://doi.org/10.1007/978-3-031-54082-0

Credits

Characters were drawn by Estelle Chauvard

Page 1: Photo by Erik van Leeuwen

Page 27: Painting by Godfrey Kneller

Page 120: Photo by Reuters/courtesy of the Archives-California Institute of Technology/Handout

Page 165: Photo by Juice Dash/Shutterstock.com

© The Author(s), under exclusive license to Springer Nature Switzerland AG 2024

A. Aftalion, *Be a Champion*, Copernicus Books,
https://doi.org/10.1007/978-3-031-54082-0

Printed in the United States
by Baker & Taylor Publisher Services